BIOLOGICAL STRUCTURE AND FUNCTION : NO. 1

BIOLOGY OF BONE

BIOLOGICAL STRUCTURE AND FUNCTION

EDITORS

R. J. HARRISON

Professor of Anatomy
University of Cambridge

R. M. H. McMINN

Professor of Anatomy
Royal College of Surgeons of England

BIOLOGY OF BONE

N. M. HANCOX

Professor of Histology and Cell Biology (Medical)
University of Liverpool

CAMBRIDGE
AT THE UNIVERSITY PRESS
1972

Published by the Syndics of the Cambridge University Press
Bentley House, 200 Euston Road, London NW1 2DB
American Branch: 32 East 57th Street, New York, N.Y.10022

© Cambridge University Press 1972

Library of Congress Catalogue Card Number: 73–169578

ISBN: 0 521 08342 7

Printed in Great Britain
at the University Printing House, Cambridge
(Brooke Crutchley, University Printer)

CONTENTS

PREFACE

Because of its unique properties, bone attracts the interest of many different kinds of investigators who, using many different techniques, add in innumerable different ways to our understanding of its structure and function. Though a monograph could (and should) be written by a physicist, a chemist, a palaeontologist, a radiologist, or an orthopaedist, this one happens to be about bone as seen by a histologist. So it is concerned primarily with structures and happenings in bone at the cellular and subcellular level. Among its many shortcomings is the omission of a number of topics which are traditionally regarded as central to the bone story. Thus, endochondral ossification has been omitted altogether, whilst the healing of fractures is given only the barest mention. This is partly because all of these topics really require (and have) monographs of their own and partly because, in the end, what to write about is a matter of personal predilection. Again, there are the deficiencies inherent in the histological approach; those seeking quantitative data are, unfortunately, likely to be disappointed.

The first section, in a way, sets the scene. It deals first with some of the properties of bone as a tissue, and then gives an outline of the development of techniques for the study of bone and the growth of ideas about its structure. There follows a description of the chemistry of the main bone constituents, then an account of how these are put together to construct bone tissue of various kinds. The first section next concludes with a short review of the evolution of bone.

For the subsequent sections, a more dynamic viewpoint is taken. The second and third deal, in turn, with the activities of the specific cells involved in the deposition and resorption of bone, and with factors involved in their control. The fourth and final section is an attempt to gather together and examine, under the heading of 'stimuli', some of the various agents, tendencies and accidents which seem to trigger off osteogenesis.

Dr Brian Boothroyd has made major contributions to this monograph. He has prepared all the electron micrographs used for illustrations; he has written the accounts of the chemistry of bone salt and of the organic constituents of bone, as well as the section dealing with calcification mechanisms.

Acknowledgements are due to Dr J. V. Occleshaw of Walton Hospital for the X-rays reproduced on Figs. 71 and 76 and to Mr G. Kidd who prepared the illustrations used on Fig. 5 after a diagram by my late colleague Dr J. Kruszynski. Dr Marshall Urist kindly supplied the sample of demineralized matrix used for Figs. 79 and 80. Thanks are due to the British Museum for the specimens from which Figs. 24–28 were prepared. Material provided by the late Dr Mary Sherman of the Oschner Clinic, New Orleans, has been of the utmost value, and was used for Figs. 57, 58, 59, 72, 73, 74 and 75. I am grateful to Dr Clifford Jones for helpful advice and for allowing me to use one of his preparations from which Fig. 45 was prepared.

Finally, I would like to express my gratitude to my colleagues and friends for their patient forbearance; especially to Mrs Brenda Brooks whose technical assistance with histological preparation and microphotography has been indispensable and to Mrs D. Godsell for her expert secretarial help.

N. M. HANCOX
February 1972

SECTION 1

HISTORICAL AND DESCRIPTIVE

1

GENERALITIES

Bone tissue is the most complex of all the building materials of the body. It is probably the last to have appeared in evolution. Its unique physical and chemical attributes mirror the diversity of its functions.

The most obvious, of course, is the skeletal one. Strong, hard bones make useful limbs. Interestingly enough, useful limbs help to maintain strong hard bones; if a limb is immobilized for any length of time, X-ray or chemical examination of its bones will reveal a loss of bone substance ('disuse atrophy' or osteoporosis), and there will be, correspondingly, loss of mechanical strength (Heaney, 1964). Similarly, mineral is withdrawn from the bones of astronauts subjected to lengthy periods of weightlessness (Mack *et al.* 1967; Hattner & McMillan, 1968; Editorial, *British Medical Journal*, 1970). A second function is that of protecting vital soft tissues; apart from brain and spinal cord, a good example is given by the bone marrow. Deep within the bones, sheltered completely from external pressure, is the haemopoietic tissue of the marrow. In this tranquil environment, newly formed blood cells can crawl from the soft haemopoietic tissue into thin-walled blood vessels and so gain access to the peripheral circulation. There is another kind of dependence between these latter tissues; it has been shown that fragments of bone marrow, transplanted to the anterior chamber of the eye (and elsewhere), are capable of inducing the newly-formed connective tissue around themselves to transform into osteogenic cells which then form plaques of bone (Scowen, 1940, 1941, 1942). Conversely, bone caused to form in the anterior chamber of the eye following transplants of osteogenic cells induces the appearance of bone marrow (Heinen, 1952).

Third, there is the role of bone in mineral metabolism. Bone matrix is impregnated throughout with calcium phosphate (in the form of crystals of hydroxyapatite) which is the principal bodily reservoir of these ions. The blood calcium concentration is maintained at a steady level in normal conditions because mineral can be mobilized from the bones to the blood under the control of the parathyroid hormone, whilst, conversely, mineral can be taken from the blood and stored in the bone matrix. It is possible that the newly-discovered thyroid hormone calcitonin (MacIntyre, 1967; Hirsch & Munson, 1968; Copp *et al.* 1969; Rasmussen & Pechet, 1970) may be involved in this latter process. Calcitonin lowers the blood calcium level. Though its mode of action is still uncertain it seems likely that in part at least there is a direct effect upon bone tissue.

Bone plays a fourth role as a trap for a variety of blood-borne ions which may exchange with calcium ions, or otherwise become incorporated in the apatite crystal lattice, or become bound to the organic matrix. In chronic lead poisoning, for example, large quantities accumulate in the bones by substituting for calcium ions in the mineral phase. Fluorine is bound in a similar manner; its presence may lead to morphological changes in the bones (Faccini, 1969). Minute quantities of fluoride (so small as to be inert in all other respects), added to drinking water, act upon the hard tissues of the teeth. Though the mechanism itself is poorly understood, the effect is to produce a dramatic reduction in the incidence of dental caries (McClure, 1970).

Radioactive substances form another group of elements, comprising for example, radium, used industrially at the beginning of the century, and atomic fission products such as radio-strontium, which 'lock' into the apatite crystal. Examples of radioisotopes which probably attach to the organic matrix are plutonium and americium (Williamson & Vaughan, 1964). Once anchored in the bone, these 'bone-seeking' isotopes irradiate their surroundings, including the nearby bone marrow. The latter may be so damaged or destroyed that red and white cell production may cease, or leukaemia may ensue. It is a sad reflection for the biologist on the times we live in that bone tissue, which evolved over 400 million years ago as a key event in the rise of the vertebrates, has become a potential liability to us.

Bone is the hard tissue from which are constructed the distinctive anatomical entities we call bones, but it is not confined to them. Extra-skeletal bone deposits are a common, normal, finding in some species; for example, leg bone tendons normally ossify in the turkey. In other species, including man, extra-skeletal ('ectopic') bone does not normally occur, but can be induced to appear under abnormal conditions. For instance, transitional epithelium is sometimes accidentally transplanted during operations which involve incision of the urinary bladder; small aggregates of epithelial cells may be disseminated in the wound area, usually in the intramuscular or sub-dermal connective tissue. Some of the epithelial nests survive; they proliferate in their new location. After about ten to fourteen days, small tumours appear. On section, these are seen to comprise a central lumen bordered by the transitional epithelium, with a connective tissue wall within which small plaques of bone arise. Extra-skeletal bone is also formed in the condition known as myositis ossificans; here, following mechanical injury, portions of muscles, usually in the forelimb, undergo ossification. The underlying changes involved in ectopic bone formation are discussed in more detail in Chapters 16 and 17.

Bone tissue is a difficult subject for histological and cytological work. The inherent problems of preparing thin sections from a rock-hard material are sometimes formidable and for many purposes it is customary to soften bone by dissolving its mineral constituents. This has been done since the earliest days of bone microscopy (see p. 7). Such 'decalcified' bone retains its organic framework, its cells, soft tissues, etc.; and if the decalcification is carried out carefully, they may remain virtually unaltered from the morphological point of view. Though still hard and tough to cut, sections can be made from decalcified bone without too much difficulty. For this reason the great majority of histological researches upon bone have, in the past, been carried out on decalcified bone. However, methods for preparing sections of whole, undecalcified bone have become more practicable, especially in the last two decades. These are discussed in more detail below. They have opened up the way to the microscopic study of the distribution and concentration of the bone mineral itself.

Bone has always been a subject of great interest to biologists in general; to anatomists, histologists, physiologists, biochemists and biophysicists; to engineers, radiologists and endocrinologists, and, in more recent times, to orthopaedic surgeons. The rise of orthopaedics first as a part of general surgery and later as a speciality in its own right has had an important influence on bone research. The practical, clinical problems encountered by orthopaedic surgeons have focused interest upon and made money available for laboratory work with bone; orthopaedic journals have provided a useful forum for the publication of research results. Recently, a specialized scientific journal devoted exclusively to work on mineralized tissues (*Calcified Tissue Research*, vol. 1, 1967), has come into being.

2

DEVELOPMENT OF METHOD
AND VIEWS

The history of histology is closely bound up with the development of microtechnique. Advances in histological knowledge have come about through the application of new methods to bring successive harvests of fresh facts and new theories. This is especially true for bone.

The pioneer microscopical observations on bone were made by the Dutch amateur Anthony van Leeuwenhoek (1632–1723). Using the simple microscope, a single, powerful biconvex lens capable of providing magnifications ranging from forty to around two hundred diameters, he produced the first drawings and descriptions of many important biological entities such as bacteria, infusoria, parasitic intestinal protozoa and red blood corpuscles. The real scientific significance of his findings was not, of course, apparent at the time, and they were thought of more as intriguing curiosities than fundamental discoveries. A good account of the man, his methods and microscopes is given by Dobell (1932).

Van Leeuwenhoek wrote to the Royal Society (of which he had been made a Fellow in 1680) about his microscopic studies on bone (1693). Essentially, he had found the solid matrix to be penetrated by four different kinds of 'tubuli'. The three largest were, in all probability, examples of what would now be known as Haversian canals, and the smallest, very likely, of osteocyte lacunae. Though visible to him in 'Shivers' taken from pieces of broken bone, he could not see these smallest tubuli in the surface of bone cut across with a sharp knife. Failure to distinguish lacunae in his sectioned bone samples cannot have been due to inadequacies in his lenses, which were, in fact, capable of resolving very much smaller objects such as bacteria, or red blood corpuscles; the fault probably lay in the lack of suitable methods for specimen preparation and of inadequate illumination technique.

At about the same time, and quite independently, Clopton Havers (1692) also identified canals running along bone. These, of course, were later named after him. He thought their function was to distribute marrow oil so as to 'mollify' the bone substance and thus keep it healthy. The original publication is not easy to obtain, but a facsimile reproduction of part of it is given by Enlow (1963).

[6]

Howship (1817) published some observations and illustrations of pathological bone, made with the aid of the solar microscope. This was a form of relatively low magnification projection microscope, utilizing sunlight to project an image which could then be drawn around, thus simplifying the accurate representation of histological appearances. Howship noticed that the longitudinal bone canals in some specimens seemed to have enlarged smoothly and symmetrically but other canals were less equally enlarged, the sides of the cavities exhibiting a rough and uneven appearance. He inferred that smooth, symmetrical enlargement was produced because the membranes of these canals became absorbing surfaces without losing their smooth, even texture. Where the rough, uneven appearances arose, the membranes not only became thicker and more vascular, but took on a granulated structure externally, where the surface of absorption acts upon the surrounding bone. Such ragged, punched out irregularities in the otherwise smooth contours of the bone canals are now known as Howship's lacunae (see Figs. 57, 58, 59 & 75). They are nowadays accepted as the microscopical hallmark of bone erosion. However, the causal relationship between lacunae and bone erosion was first illustrated and commented upon, by Tomes & de Morgan only some sixteen years later (1853).

Purkinje first wrote about osteocyte lacunae which he called 'bone corpuscles'. The name 'lacuna' was suggested by Todd & Bowman (1845) who also introduced the concept of the 'Haversian system' as comprising a central, vascular canal together with an endowment of surrounding, circumferential, concentrically-arranged bone lamellae.

Quekett (1846) made comparative studies of the structure of bone in various species. His illustrations are surprisingly modern-looking, but interstitial lamellae between Haversian systems find no mention in the textual description. He believed that the material in which the lacunae are set, 'the true bony substance', or 'ossific substance' as distinct from the 'earthy matter' (i.e. the bone material) consisted of small globules. In all probability the 'globules' were an optical artefact created by the imperfections of his microscope, for, with optical systems of relatively low resolution, circular diffraction patterns tend to make small objects look much larger than they really are. What is possibly the earliest reference to histological decalcification of bone is contained in this paper; Quekett mentions the deliberate use of muriatic acid to remove the earthy matter.

Goodsir & Goodsir (1845) refer to the bone corpuscles and their canaliculi. They noted that the former contained nucleated cells. They studied both adult and growing bone and they illustrated, for the first time, a close morphological association between newly-formed matrix and osteogenic cells, even though the technical means available to them were inadequate to demonstrate that these cells possessed any specific characteristics; these same cells were to receive their name of osteoblast from Gegenbauer some years later. Anyone interested in reading the Goodsirs' article and unable to procure it easily in the original will find it reprinted in *Clinical Orthopaedics* vol. 40, 1965.

Advances in biological microtechnique, which were to make possible the study of histology as we know it today, had by this time started. Although specimen preparation remained relatively crude, without stains or methods for cutting thin sections, the optical performance of the microscopes benefited from the introduction of achromatic corrections for objectives and the use of achromatic substage condensers. The work of Tomes & de Morgan (1853) reflects the improved quality of the microscope; their illustrations indicate a greater image crispness partly due to decreased depth of focus and comparative lack of the globular diffraction effects which troubled the earlier microscopists; and partly to improved specimen preparation. They made several profoundly important new observations. For instance, they demonstrated absorption lacunae in transected Haversian canals and explained how their presence signifies that in the normal course of events old bone is removed and new laid down in its place. This provided microscopical evidence to support the ancient suspicion that, in the animal economy, the old and the worn out are removed and replaced by fresh material. They found that bone remodelling (as we call it today) occurred at all ages, though it diminished with increasing age; that some osteocyte lacunae became 'in great part filled up with solid matter', as reported over 100 years later by Jowsey (1960) using an altogether very much more sophisticated technique. They illustrated interstitial lamellae and explained them as 'remaining parts of Haversian systems, the larger portions of which had been removed by absorption'. They made important observations on the mechanism of bone removal. They understood that a special cellular activity must be involved; this is shown by the following extract from their paper given at length because of its unique importance:

During the present winter it became necessary to remove a portion of the femur which protruded from a stump six weeks after the removal of the limb. From the medullary cavity a granulating mass projected and covered the surface of the bone left by the saw, and as the bone was rapidly wasting from the inner or medullary surface, we had in this specimen a favourable opportunity of examining the tissue which lay in immediate contact with the surface of the wasting bone. On cutting through this piece of femur in its length with a very fine jeweller's saw, it was found that a dense, pale pink tissue lay in contact with the inner surface of the bone, which was hollowed with numerous minute cavities, into which the soft tissue accurately fitted, but from which it could be detached without tearing. The outer surface of the bone had been deprived of membrane many days before its removal from the limb.

The examination of the tissue thus closely applied to the fast wasting bone, offered as favourable an opportunity for learning something of the means by which absorption is effected as we could reasonably expect to obtain, the more so since the outer surface having been for some time exposed and covered only by dried periosteum, the actions had been confined to the inner surface of the bone. A careful examination showed that the surface of this tissue was composed of minutely granular nucleated cells, which lay in an exact ratio with its diminution. What the bone lost in bulk the cells gained, the cellular mass presenting a perfect cast of the surface of the bone, suggesting to the mind that the soft was growing at the cost of the hard tissue, or at all events that the former was instrumental in the removal of the latter. The cellular mass was tolerably vascular, but the vessels did not reach the surface in contact with the bone; hence they could not be regarded as having any immediate action in the process of absorption. Section of the bone showed that the medullary cavity had been greatly enlarged by absorption, and no doubt had sufficient time been allowed the femur at that part would have been reduced to a thin scale.

Though their technical resources were insufficient for them to make a cytological characterization of the bone-absorbing cell itself, the osteo-clast, they remark: 'Although the process of absorption has not, and probably cannot, be seen in actual operation, yet a consideration of the relative position of the increasing and wasting parts, and of their condi-tions, will, in the authors' opinion, justify the conclusion that the bone is removed through the agency of cells'. Osteoclasts were first described and christened by Kölliker in 1873.

Tomes and de Morgan also studied the growth of bone. They illustrate 'osteal' cells in foetal lamb and human material, and identified them, as well, in adult samples, as associated with bone formation. These cells were obviously osteoblasts. Unfortunately, Tomes and de Morgan did not believe in the presence of fibres in bone matrix and therefore their views on its construction and on endochondral ossification make strange and difficult reading today.

During the latter half of the nineteenth century, advances in biological microtechnique continued rapidly; not only microscopes but histological preparations improved greatly. For instance carmine was introduced by Gerlach in 1858 as a background colourant or stain and helped to bring fresh detail to light; paraffin embedding, by Klebs, in 1869, made thin sectioning feasible; haematoxylin was brought into use as a nuclear stain by Boehmer in 1865. A short account of the history of biological staining is given by Mann (1902). Oil immersion lenses for higher resolution and magnification were mass produced by Leitz and became available 'off the shelf' around 1878. By 1880 there were several excellent atlases of histology, often adorned with beautiful coloured lithographic plates.

It is very interesting to examine the text and illustrations of bone in a book of that period (for instance, Klein & Noble Smith, 1880) in the light of present-day knowledge. In many respects, they cannot be bettered, but naturally, in others, we now know a great deal more. One major deficiency in knowledge concerned the organic matrix. Though the presence of bone lamellae had been recognized for many years, the fact that they contain collagenous fibres, disposed in a specific way, was still to be established. However, in 1883 Klein remarked that the 'matrix of the osseous substance is dense fibrous connective tissue, i.e. a tissue yielding gelatin on boiling'.

Prominent among the techniques used in the study of the constitution and disposition of bone lamellae was the use of the polarizing microscope, particularly by v. Ebner (1875, 1894) and Gebhardt (1906). Ranvier, in the second edition of his *Traité Technique d'Histologie* (1889) gives an illustration of lamellae as seen in polarized light. He suggests (rather acidly) that v. Ebner's work was inspired by remarks made in the first edition of the *Traité*. With polarized light, a bone section gives a characteristic pattern of dark and light areas. That this is due mainly, but not entirely, to its constituent collagen fibre bundles follows from the fact that sections of bone tissue which have been decalcified beforehand behave in practically the same way in polarized light as sections from whole bone.

In the polarizing microscope, collagen fibres are anisotropic. When viewed through crossed nicols, fibres cut more or less transversely will fail to transmit light and will therefore appear black, in all azimuths. Fibres cut longitudinally will, in contrast, appear brightly-illuminated except where they happen to be parallel with the plane of polarizer and analyser. There, they will appear dark and so give rise to the characteristic Maltese cross pattern seen at its clearest with well-developed concentric lamellae (Fig. 15).

Microscopic study of bone has always been technically difficult. To begin with, it is plainly impossible to cut thin sections of whole bone by conventional methods because it is so hard that the edge of an ordinary steel microtome knife would be chipped or turned over. The choice has to be made between four alternatives. First, it is possible, after fixation, in the normal way, to decalcify bone, the stony hard bone mineral being removed in a variety of ways such as by solution in mineral or organic acids, or by chelation for instance with ethylenediaminetetracetic acid ('EDTA' or 'versene'). The remaining tissue constituents stay more or less unaltered if decalcification is done carefully and the sample can then be dehydrated, cleared, embedded and sectioned in the usual way. Actually, even after decalcification, bone is quite difficult to section by ordinary procedures; its extensive collagen content becomes tougher and harder during processing. In sections of decalcified bone, the collagen ground-substance framework, and the osteocytes with their processes together with their surrounding lacunae and canaliculi, are often reasonably well preserved, and of course the other cells and soft tissue constituents, such as marrow, fat and vessels, are also more or less unaltered in appearance.

The next alternative retains the bone salt but the soft tissues suffer; this is to cut out a thin slice of bone and to grind it down as thin as possible, using a fine abrasive. After washing, such preparations are often dried. The bone salt is preserved intact. When examined in transmitted light, lacunae and canaliculi appear black, because they contain air. The main drawback to ground sections, apart from the fact that they are inevitably rather thick, is that soft tissues and cells are lost, and so there is no prospect of studying the relationships of cells to matrix.

The third alternative is to use one of the techniques which have become available in recent years for the sectioning of whole, intact bone. They stem originally from critical studies, with the light microscope, on the location of bone salt in normal osteogenesis and fracture healing (Bloom & Bloom, 1940; McLean & Bloom, 1940, 1941; Urist & McClean, 1941). The first difficulty which had to be overcome for this work was to prevent artefactual loss of bone salt through inadvertent decalcification. To prevent this happening, the pH of all watery solutions used for processing and staining was maintained at a sufficiently high level to prevent solution or leaching out of mineral from the tissue samples, or from the sections cut from them subsequently. The same problem naturally confronts the present-day electron microscopist in an even more acute form (Boothroyd, 1964). A second difficulty was that of obtaining good sections and for this purpose it was necessary to use a hard embedding medium, celloidin, to support the tissue. These methods can give very beautiful results, and, with appropriate stain techniques, give valuable information about both hard and soft constituents of bone (Fig. 32). On the other hand, they are subject to the limitation that generally only comparatively sparsely-calcified samples can be cut without damage. The methods worked up by Bloom and McLean have, in recent years, been brought to a further stage of development to meet the needs of modern research on the distribution of the actual bone salt.

The requirement for work along these lines is for thin undecalcified sections, from even the most fully mineralized and densest bone samples. To provide such sections, two technical improvements have been introduced. On the one hand, use is made of substances much harder than those used traditionally for embedding, so that the sample to be cut is incorporated into a block of supporting material as hard (or nearly so) as the bone itself. Polymer plastics have been widely used, and, lately, low melting-point metal alloy ('Cerrolow 17'), to hold the samples in place against the cutter and to help to prevent the sections from breaking up. On the other hand, specially shaped and hardened steel microtome knives have been evolved; and also rotating cutters, basically resembling the pattern used in metal milling machines (Jowsey, 1955), can be used. A further refinement is the use of the diamond 'knife' for preparing ultra-

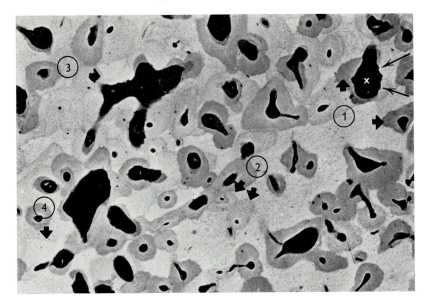

Fig. 1. Microradiograph of human bone. Magnification, × 120. Non-mineralized areas (i.e. vascular channels) appear black; most heavily mineralized are palest. At least four levels of increasing density (numbered arrows) can be distinguished. Generally speaking, the channels are surrounded by least mineralized matrix. Largest channels show evidence of bone resorption. The outline of channel 'x', for instance, is irregular, with Howship lacuna on its left margin; the contour also passes through older, densely mineralized bone (oblique arrows).
Preparation kindly supplied by Dr A. Ferguson.

thin sections for electron microscopy. This is shaped like a tiny plane iron, the cutting edge being ground from an industrial diamond. Cost and other factors limit the length of the cutting edge to a millimetre or two, which is adequate to cope with the minute samples used for electron microscopy. This brief account of technical progress in the microscopical study of bone would be incomplete without some mention of three comparatively recent innovations for studying bone mineral as preserved intact in such sections.

First, work can be carried out on the distribution and relative concentrations of the mineral by means of microradiography. This entails preparing soft X-ray photographs of undecalcified sections; heavily-calcified areas, being radiopaque, come out black in the final picture. Less calcified areas appear in shades of grey whilst unmineralized parts are white (Fig. 1). This ability to discriminate between different densities of calcification in different parts of a section has provided the basis for new investigations on the dynamics of the mineral component of bone.

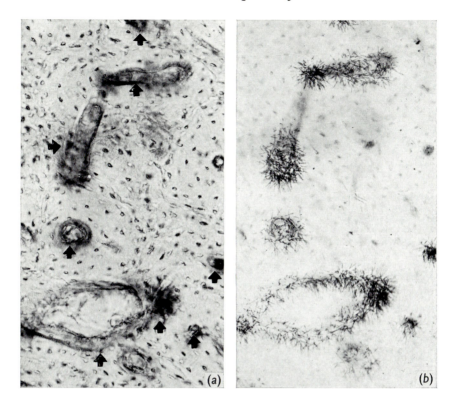

Fig. 2. Autoradiograph using americium. Magnification, × 155.

(*a*) Undecalcified section of dog bone. Arrows point to vascular canals, around which bone is normally in active process of deposition or resorption. The associated concentric bone lamellae can be recognized around one or two of the canals. The ubiquitous small spaces are osteocyte lacunae.

(*b*) Same field as (*a*), objective raised to focus on overlying photographic emulsion. Note the dark, linear tracts, which lie over the vascular spaces seen in (*a*). The isotope is evidently present in the bone matrix immediately around the canals; none is located deeply within bone.

Preparations kindly supplied by Dame Janet Vaughan.

Second, the sites where exchange of mineral is taking place may be studied with 'labelled' tracers which lock into such areas. Several kinds of label are available; one of the most popular is tetracycline which is more or less harmless and which can be identified in undecalcified bone sections from its specific fluorescence in ultraviolet light (Frost, 1969).

Other labels include the bone-seeking radioisotopes whose distribution can be mapped out by autoradiography (Rogers, 1967). With this technique the experimental section is prepared from the bone of an animal dosed

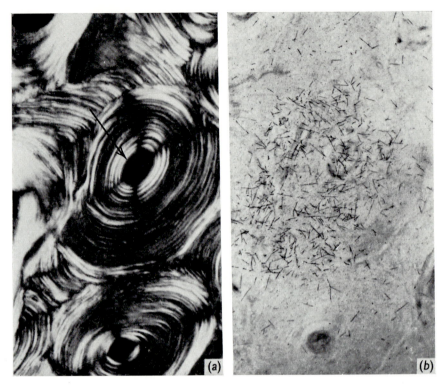

Fig. 3. Autoradiograph from bone of patient dying from radium poisoning. Magnification, × 155.

(*a*) Undecalcified section of bone, viewed in polarized light. The dark space (arrowed) in the centre of the field is the vascular canal of a Haversian system whose concentric lamellae can be seen around it.

(*b*) Same field as (*a*), objective raised to focus on overlying photographic emulsion. The dark linear tracts correspond to radiation emitted by the radium. Note that the radium has no preferential distribution with this Haversian system; it is located more or less evenly amongst the lamellae.

Preparations kindly supplied by Dame Janet Vaughan.

with an isotope or exposed to atomic fall-out. The section is then mounted on a slide and coated, in the dark, with a thin film of photographic emulsion. The resultant sandwich (that is, slide on one side, section in the middle and emulsion on the other) is put aside in the dark. Radiation from any 'hot spots' in the section will travel outwards, and will collide with silver halide particles in the film above, and reduce them. After a suitable interval, the whole preparation is taken through photographic development, and the reduced silver grains overlying the hot spots will appear black. The emulsion itself is transparent and the section beneath can be stained in the ordinary way to bring out structural details. In the final preparation, therefore, one focuses down first upon the black silver grains in the plane of the emulsion, then on to the stained section beneath, and so the morphological identity of any isotope-labelled areas can be established (Figs. 2*a* & *b*, 3*a* & *b*).

A further example of a bone mineral label is provided by lead (Vincent, 1957; Hong *et al.* 1968*a*, *b*; Schneider, 1968). After injections of a soluble salt, lead is taken up from blood or tissue fluid and 'locks' into apatite crystals newly deposited in the matrix. Samples of bone are then removed, fixed, and carefully decalcified in dilute acid in the presence of excess hydrogen sulphide. Under these conditions, the lead is converted to insoluble lead sulphide which remains behind in the gradually softening tissue. Finally the bone can be embedded (in gelatin) and sections cut on the freezing or cryostat microtome. The lead sulphide strata can be visualized by means of a gold chloride method. Though the decalcified specimens can be sectioned without serious difficulty, technical problems with this method tend to undermine the validity of the localizations if resolution is pressed too far. As with tetracycline and the infinitely more arduous radiotracer methods, a single injection of lead can give apparently reliable histological information as to sites in the body where bone deposition happens to be in progress at any one time (Fig. 4*a* & *b*). Two or more injections given at known intervals will each produce a labelled 'line' in bone and measurement of the inter-line distances can give quantitative information about the actual amounts of bone deposited per unit of time.

Apart from its mineral constituents, the organic framework of bone can also be coloured or 'tagged' as it is laid down. The bones of animals fed with madder look pink because they contain the dye alizarin which combines with newly-deposited bone matrix to form a relatively insoluble compound. Perhaps the best-known early work with madder was carried out by Belchier whose article is reprinted in full in *Clinical Orthopaedics*, vol. 40, 1965. Vilmann has recently (1969) found that the active principle colours resorbing, as well as newly deposited, bone. More recently, dyes of the Procion series have been used as markers of newly deposited matrix (Prescot *et al.* 1968) but are probably very toxic.

The third kind of recent microscopical work upon bone is concerned with the analysis of its inorganic constituents by means of the electron-probe X-ray scanning microscope. The principle involved here is to focus a beam of highly accelerated electrons on a small area of the surface of a prepared specimen. The target atoms absorb some energy from the electron beam and are raised to higher energy states; when allowed to return to their normal state, they emit a 'flash' of X-rays whose wavelengths are characteristic for the atoms involved and whose intensity is proportional to their concentration. The emission of X-rays can be analysed

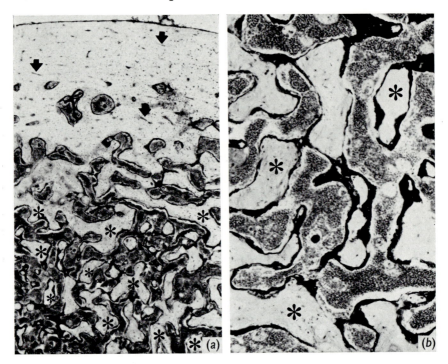

Fig. 4. Part of cross-section of femur of laying hen which had been injected 48 h previously with lead citrate; sites of lead deposition revealed as black lines and aggregates by the technique of Hong *et al.* (1968).

(*a*) Low-power (magnification, × 40) view. Unlabelled bone appears white. The upper part of the section comprises the outer cortical bone in which a few vascular canals run (vertical arrows). The lower part contains the interlacing trabeculae of the medullary bone (asterisks) which of course emanate from the cortex. Between them lie blood vessels and marrow (dark). Even at low magnification, black lead lines can be distinguished on the edges of many of the trabeculae.

(*b*) Higher power (magnification, × 120) view of part of the medullary bone. The bone trabeculae (e.g. asterisks) appear white; the tissue between is blood and marrow. The intense black areas on trabecular edges correspond to sites of lead accumulation and, hence, to bone deposited since the time of the injection 48 h previously.

both qualitatively and quantitatively, and thus information about the nature, proportions and linkages of the elements in the bone mineral (Mellors, 1966). The same principle has been applied recently to the calcified spicules of sponges (Jones & James, 1969) and to the radular teeth of the limpet (Runham *et al.* 1969).

The methods of tissue culture have been applied to bone with interesting (and, in some instances, crucial) results. These experiments will be described at appropriate points in later chapters; they have been most useful, perhaps, in work undertaken to clarify the specific roles of the cells customarily native to bone, and the effect upon their functions of various hormones.

Finally, of course, the point must be made that research on bone is by no means the exclusive province of the biologist whether he uses the methods of histology, pathology, or biochemistry. The physicist, mathematician, engineer; the palaeontologist, orthopaedist and dentist all contribute to our knowledge. The real problem is one of communication.

3

VARIETIES OF BONE

In man and the higher mammals, bone is built up of four basic constituents which are employed in the construction of bone tissue of two varieties, namely woven and lamellar bone. The individual properties of these structural components will be discussed fully in the next chapter; however, they must be reviewed briefly at this point, in order that the main histological differences between the two bone varieties may be explained.

The first of the constituents is a protein scaffold, forming the main bulk of bone. With the light microscope this is seen to consist of collagenous fibres and of course in the electron microscope the fibres are observed to comprise bundles of very much finer fibrils carrying the distinctive periodic cross-banding of about 640 Å. Second, there is the ground substance, an amorphous material which is apparently structureless in both the light and electron microscopes; chemically, it consists of protein–carbohydrate complex. It is believed to permeate the collagenous bone matrix, enveloping the fibres. The third component is the crystalline bone mineral, insoluble at body pH, which confers upon bone its stony hardness. The fourth component is built up of the osteocytes (the bone-inhabiting cells), each of which occupies its own cavity, or lacuna, within the matrix. Osteocytes possess numerous long, cytoplasmic processes which travel through the matrix in minute tunnels, the canaliculi, and anastomose with each other. The relative quantities of these constituents present in bone tissue, and their histological arrangement, vary.

For example, in some species, osteocytes are completely absent. There are none in many *teleostii* (Kölliker, 1859; Moss, 1963). This complete lack of cells within their bone does not seem to have retarded the evolution of these creatures, nor to have affected the usefulness of their bones in any way. In other *teleostii* (Moss, 1965) some parts of a particular bone may be completely acellular whilst osteocytes are present in neighbouring areas. Species differences may exist between the shapes and sizes of lacunae and canaliculi (Moodie, 1926; Crawford, 1940). Full details of the facts are set out in the reviews of Enlow & Brown (1956, 1957, 1958) and in the reports of Moss (1961, 1962, 1963, 1965) and of Suzuki (1963).

In man and the higher mammals the proportions and histological arrangement of the bone constituents fall into two chief combinations,

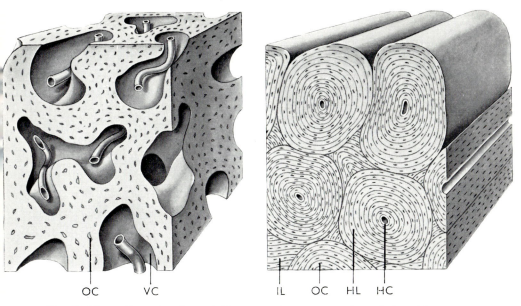

OC   VC   IL   OC   HL   HC

Fig. 5. Three dimensional, semi-diagrammatic representation of the differences between woven and lamellar bone. Woven bone, on the left, is permeated by relatively large and tortuous channels (VC) containing fat, blood vessels, connective tissue. Osteocytes (OC) are distributed haphazardly in the bone matrix. Lamellar bone is on the right; it has interstitial (IL) and Haversian (or concentric) lamellae (HL). Haversian canals (HC) convey blood vessels. Osteocytes (OC) are arranged in regular, concentric order.

producing, respectively, the so-called *woven* or *coarse-fibre*, and *lamellar* or *fine-fibre* bone. Both woven and lamellar may enclose only sparse small vascular channels, giving rise to *compact* (dense) bone; or they may be permeated by many large vascular spaces whose aggregate volume equals or exceeds that of the bone matrix itself, forming *cancellous* (spongy) bone. The principal features of woven and lamellar bone are described below.

WOVEN BONE

In its common form, a three-dimensional block of this material may be compared with gruyère cheese; both consist of large irregular sac-like spaces separated by 'walls' of substance (matrix) of varying thickness (Fig. 5). However, when seen in the two-dimensional view presented by histological section, the 'walls' are known as 'trabeculae'. If the section plane happens to provide a plan view, the sectioned 'walls' appear as broad sheets or planes (Fig. 8) but as narrow 'trabeculae' when cut in side elevation view (Figs. 6 & 7).

Within the trabeculae, the collagenous fibre bundles are relatively

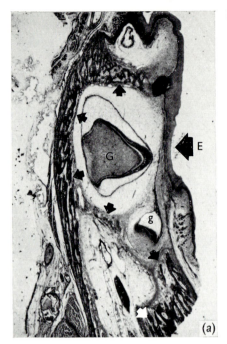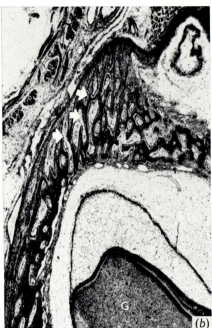

(a)

(b)

Fig. 6. Section of part of decalcified embryonic lower jaw. Masson trichrome stain.
(*a*) Magnification, × 10. A deciduous (G) and a permanent tooth (g) germ lie below the gum epithelium (arrow, E). They are surrounded by woven bone trabeculae (arrows) of the developing mandible.
(*b*) Higher magnification (× 25) of a part of preceding. The interlacing pattern of the trabeculae (arrows) characterizes the two-dimensional view provided by sections of woven bone.

coarse. Characteristically, they interweave, running in all spatial directions, and so make up a material comparable to carpet underfelt. Therefore in histological section, the collagen fibres will be cut in all sorts of section planes (Figs. 9 & 10). The lack of a strongly-ordered arrangement of the fibres is reflected in the relatively undramatic appearances produced by polarized light in woven as compared to lamellar bone.

Another important histological criterion of woven bone is provided by the osteocytes. They are scattered more or less at random in the matrix, and quite without orientation in respect to the vascular channels (Figs. 5, 7 & 8). The canaliculi which radiate out from the lacunae are probably rather scantier than in lamellar bone, but they interconnect in a complex way and lead out to the vascular channels (Fig. 11). The seemingly random orientation of fibres and osteocytes in woven bone contrasts very sharply with the exquisite regular minute orderliness of lamellar bone.

There are probably differences in the properties of the ground substance

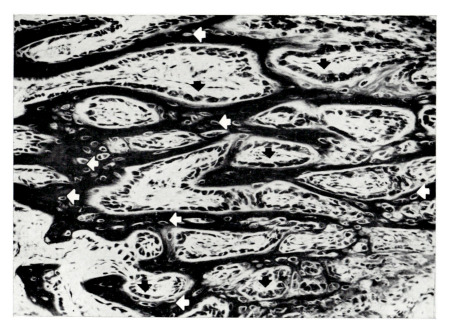

Fig. 7. Higher magnification (× 120) of part of preceding. The darkly-stained woven bone trabeculae surround irregular spaces which contain blood vessels and connective tissue. Lacunae appear as small oval or round spaces, scattered haphazardly within the matrix (horizontal arrows). They contain osteocytes. The darkly staining cells abutting directly upon the edges of the trabeculae (vertical arrows) are osteoblasts.

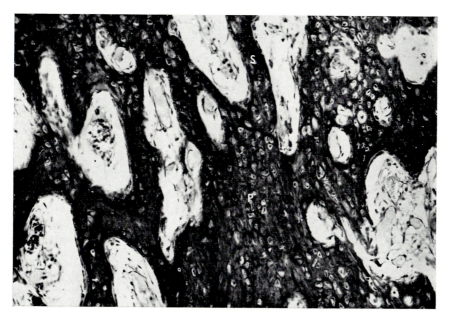

Fig. 8. Woven bone in the adult: section of decalcified human fracture callus. Masson trichrome stain. Magnification, × 140. Relatively large vascular spaces are surrounded by trabeculae of woven bone. The latter are seen both in plan and side elevation view ('S' and 'P' respectively). Osteocyte lacunae have a random distribution.

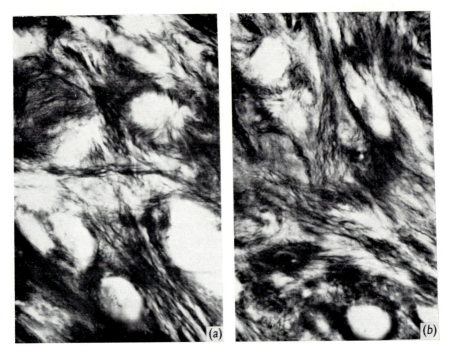

Fig. 9. Woven bone from human fracture callus. Decalcified, and sections stained by Wilder's silver method for fibres. Magnification, × 1000. In both (*a*) and (*b*) the roughly round, empty spaces are osteocyte lacunae. Collagen fibres appear black, and run in all directions; they are most easily seen when their plane is approximately the same as that of the section. Shorter lengths of obliquely-sectioned fibres can be identified here and there. The rather wavy course of the fibres is apparent. It is generally believed that they repose in a bed of transparent protein–polysaccharide, the ground substance.

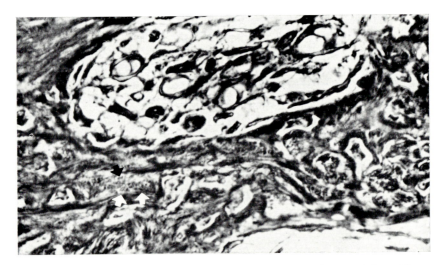

Fig. 10. Woven bone from human fracture callus. Gomori's silver technique. Phase-contrast microscopy; magnification, × 1000. Optical conditions were arranged with the object of demonstrating collagen fibres running at right angles to the section plane; they can be identified in several areas as groups of small, round, dark profiles (arrows).

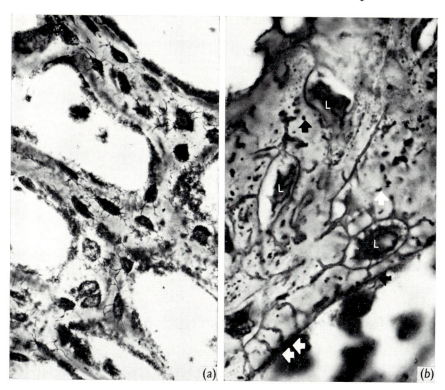

Fig. 11. (*a*) Embryonic woven bone, decalcified and ester wax embedded. Section stained by Schmorl's picrothionine method. Magnification, ×480. The large, empty spaces are vascular channels; they are surrounded by trabeculae of woven bone. Osteocyte lacunae and canaliculi are distributed haphazardly in the bone.

(*b*) Woven bone from human fracture callus. Haematoxylin and eosin. Magnification, ×1200. The outlines of several lacunae can be seen; they contain shrunken remains of osteocytes (L). Canaliculi can be seen running in various oblique and longitudinal planes; others are visible in transectional view (vertical arrows). Canaliculi can be seen to open on the bone edge (horizontal arrows).

of woven and lamellar bone. It is thought to be more reactive in woven bone; this is manifested by a more vigorous reaction with histochemical test for mucopolysaccharides, and a more pronounced metachromasia with dyes such as toluidine blue. The significance of this behaviour is discussed more fully in Chapter 7.

It seems true that (with one important exception) bone is always deposited initially as woven bone. Thus, in the embryo, the first bone tissue generated in developing bones is of woven type; and in post-foetal osteogenesis, such as fracture healing, ectopic ossifications, or osteosarcoma, new bone is deposited in woven form. Woven bone, however, less highly-organized than lamellar bone, is a relatively ephemeral material and appears to have built-in obsolescence. It never endures for long. It is sooner or later resorbed and its place is taken by lamellar bone. For instance, in the case of the human animal, the limb long bones at birth are still constructed of woven matrix but its replacement by lamellar bone soon commences; by the second or third year, replacement is finally completed. The ways in which this apparently unavoidable substitution takes place, and the factors which determine its onset are reviewed in more detail in Chapter 11. Lamellar bone is of course itself normally 'turned over' by the opposing processes of resorption and replacement. The exception referred to above applies here. Lamellar bone, resorbed in the normal course of the structural renewal of the bones, is replaced *ab initio* by lamellar, not by woven bone.

Woven bone commonly produces a three-dimensional continuum enclosing large, irregular cavities, as described above (Fig. 5). The spaces, of course, contain vascular and connective tissue elements. *Cancellous* woven bone is characteristically formed in the repair tissue ('callus') around a fracture site and in the developing mandible and other bones. Dense or *compact* woven bone is formed in the diaphysial areas of developing long bones.

Though the description of the matrix of woven bone which has just been given fits most samples, there is nevertheless some deviation around a mean, as might be expected in a biological context. For instance, in fracture callus a tissue is often encountered whose histological structure seems to be intermediate between bone and cartilage, the so-called 'chondro-osteoid'. Most of its cells lack processes, thus resembling chondrocytes, but mingled with these are other cells with the processes typical of osteocytes. Chondro-osteoid seems to be a transitional material between woven bone and cartilage, just as elastic- and fibro-cartilage appear to form a bridge between cartilage and connective tissue. It is interesting that osteogenic cells in tissue culture sometimes produce cartilage instead of bone (Fell, 1933).

In summary it can be said that woven bone comprises a matrix whose coarse collagen fibres interweave, and in which the osteocytes are distributed more or less at random. It is a provisional material, ultimately replaced by the more highly organized lamellar bone.

LAMELLAR BONE

There are two major differences between woven and lamellar bone (Fig. 5). In the latter, the collagen fibres are finer and organized into unit layers or sheets, the lamellae; whilst the osteocytes are distributed in a more regular pattern, which bears a definite relationship to the vascular channels.

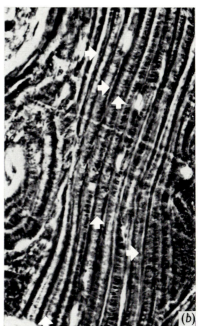

Fig. 12. Human lamellar bone, celloidin embedded. Phase contrast.

(*a*) Magnification, × 125. The sample shows evidence of remodelling; the 'sawn off' appearance of the top edge and the occurrence there of several osteoclasts (vertical arrows) indicate that bone was being removed. At the lower edge, the presence of osteoblasts and the smooth contour suggest that bone deposition was proceeding.

Two transected Haversian canals are labelled (C & c); their associated concentric lamellae are clearly seen. Above and to the right of C can be seen a number of lamellae which represent the remains of a former complete Haversian system. The position which was formerly occupied by its central canal (oblique arrow) is now occupied by lamellae belonging to C. To the right of C and c can be seen a stack of lamellae running more or less parallel to each other (between horizontal arrows).

(*b*) Higher magnification (× 400) of some of the parallel lamellae seen in (*a*). The diagonal white stripe is an artefact. Individual lamella can be distinguished. The collagen fibre orientation is the same in each lamellae but differs between neighbours. For example, profiles of groups of fibres cut transversely can be distinguished in some lamellae (vertical arrows) whilst in alternate lamellae the direction is more or less longitudinal (horizontal arrows) though individual fibres cannot be resolved in this photograph.

A lamella, a distinct, sheet-like layer of bone matrix, contains fine collagen fibres running more or less parallel to one another. As in woven bone, the structureless ground substance penetrates between the fibres, and bone salt crystals permeate the combination. The bone tissue is built up of numbers of lamellae, arranged in three main ways.

Fig. 13. Frozen section of decalcified lamellar bone. Schmorl's picrothionine method; photographed in partially polarized light. Magnification, × 120. Lacunae and canaliculi appear black; the lamellae stand out clearly in some areas because of their birefringence. In addition to the layers of concentric lamellae around Haversian canals, groups of interstitial lamellae are present (IL).

Firstly, the lamellae may lie stacked more or less flat one upon another (Fig. 12). Secondly, they may be arranged, like the skins of an onion, more or less concentrically around a central vascular 'Haversian' canal (Figs. 12, 13, 14 & 15). Such a vascular canal, together with its associated concentric lamellae, constitutes a 'Haversian system' or 'osteone'. The three-dimensional anatomy of Haversian systems was discussed by Cohen & Harris (1958). In a third category are the lamellae which are interposed between the Haversian systems proper, the so-called 'interstitial lamellae' (Figs. 13, 15 & 18). As will be discussed more fully in Chapter 11, these interstitial lamellae are, in fact, the last remnants of previous osteones, the remainder of which have been resorbed in the course of the normal turn-over of the bone substance; the space formerly occupied by the resorbed bone has subsequently been taken by the complete systems deposited later.

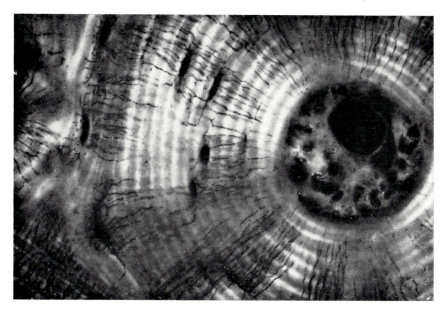

Fig. 14. Same preparation as in preceding, higher magnification (× 800). The alternation of fibre direction in the lamellae is reflected in their relative brightness or darkness. The osteocyte lacunae and their canaliculi are clearly seen.

Fig. 15. Ground section of compact lamellar bone photographed in polarized light (crossed nicols). Magnification, × 160. The characteristic Maltese crosses, generated by the concentric lamellae of Haversian systems, are visible. Interstitial lamellae are also present. Some (IL) happen to lie in the azimuths in which they are birefringent. Osteocyte lacunae can also be identified.

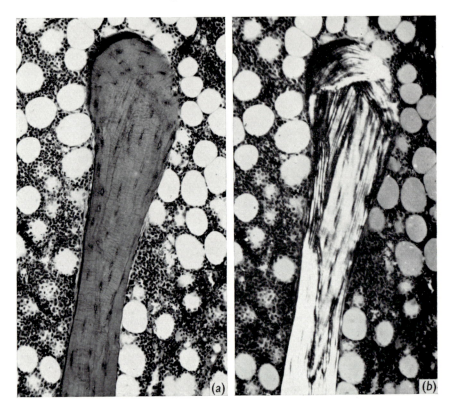

Fig. 16. Decalcified human bone. Stained haematoxylin and eosin. Magnification, ×120. A trabeculum of cancellous, lamellar bone is surrounded by bone marrow.
   (a) Normal transmitted light; note general appearance of the bone.
   (b) Same field; polarized light. The various groups of lamellae are seen distinctly.

The reason why lamellae stand out from each other derives from the orientation of their collagen fibres; though the fibres in any one lamella run parallel to each other, the fibre direction in adjoining lamellae is quite different. The lamellae can be brought to light in the microscope either by staining or by using the phase-contrast (Fig. 12) or polarizing microscopes (Figs. 13, 14 & 15). With nicols completely crossed, the typical Maltese crosses are superimposed on the Haversian systems; when the nicols are rotated a little, the lamellae appear as alternating dark and light rings corresponding to fibre axes parallel and at right angles to the section plane (Figs. 13, 14 & 15). In cancellous bone, the trabeculae are built of lamellae running more or less parallel. Polarized light brings out the lamellae very distinctly (Figs. 16 & 17). Transmission electron microscopy confirms that the collagen fibre orientation is more or less the same in each lamella (for

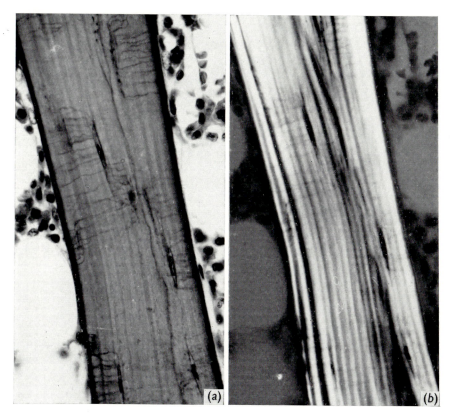

Fig. 17. Decalcified human lamellar bone. Stained haematoxylin and eosin. Magnification, × 480.

(*a*) Lacunae and canaliculi show quite clearly in normal illumination.

(*b*) With polarized light, the individual lamellae are distinctly seen. Compare Fig. 27*b*.

example, Cooper *et al.* 1966). On the other hand, studies with the scanning electron microscope (Boyde & Hobdell, 1968, 1969) suggest that some degree of anastomosis takes place between the fibres of adjacent lamellae. It is accepted at present that in all lamellae the fibres run obliquely in relation to the long axis of the Haversian canals rather than truly parallel or at right angles.

The second main difference between woven and lamellar bone derives from the arrangement of the osteocyte lacunae. In lamellar bone, they are laid out in a regular, orderly pattern which, in the case of the concentric and interstitial lamellae, can be seen at once as related to vascular channels (Figs. 13, 14 & 18). From the osteocyte lacunae, the canaliculi radiate outwards and inwards and interconnect laterally with almost unbelievable

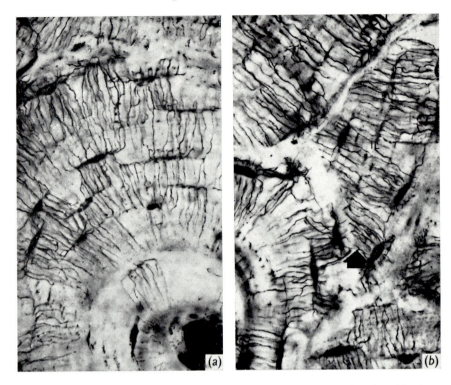

Fig. 18. Frozen section of decalcified lamellar bone. Schmorl's picrothionine. Magnification, × 480.

(*a*) Part of a Haversian system with canal at bottom right. Osteocyte lacunae and canaliculi are seen.

(*b*) A group of interstitial lamellae (IL) are surrounded by lamellae which belong to other Haversian systems. Their canaliculi do not 'cross', for the most part, but there seems to be a connection where arrowed.

elaboration. The bone matrix is so riddled with countless thousands of these tiny canals that the arrangement of osteocyte lacunae and canaliculi gives one of the most striking and beautiful of all histological images. The canaliculi belonging to one Haversian system do not often connect up with those of neighbouring systems (Fig. 18) though very occasionally they undoubtedly do so. Similarly, as would be expected, the canaliculi of adjacent interstitial lamellae generally keep to themselves (Fig. 18*b*) but this, too, is not invariably the case. In many textbook diagrams the lacunae are shown as located between adjoining lamellae but, as can easily be seen in reasonably good preparations, they are as often within as between lamellae. There are in any case many more layers of lamellae than 'layers' of osteocytes (Fig. 14). Using the scanning electron microscope, Boyde &

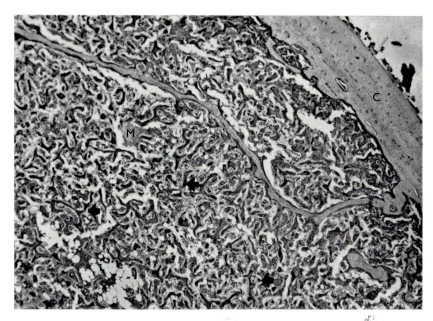

Fig. 19. Transection of femur of laying hen showing the cortical bone (C) and medullary bone trabeculae (M; arrows). The section had been stained by the PAS technique; the trabeculae of medullary bone were much more reactive and this is reflected in their darker colour in the photograph. Blood and myeloid tissue lie between the medullary bone trabeculae. Compare Fig. 4. Magnification, × 150.

Hobdell (1969) were unable to find 'any specific location of lacunae with respect to lamellae'.

The ground substance in lamellar bone appears generally to be less reactive than in woven bone, as judged by its behaviour with histochemical tests for protein–polysaccharide. For example, it is thought that with the periodic acid–Schiff (PAS) technique the intensity of the colour increases and shifts towards the purple end of the reaction spectrum as the ground substance becomes more depolymerized, and therefore more reactive. A good example of differences in bone reactivity can be seen in a cross-section of a long bone of a laying fowl. The outer shell or cortex is made of a lamellar type of bone; from its inner border, trabeculae of woven bone, the so-called 'medullary bone', extend towards the centre of the shaft. This medullary bone is thought to be turned over very rapidly indeed, being resorbed to provide calcium for the egg shell and being laid down as a mineral reservoir during the laying cycle. When stained by the Hotchkiss procedure (Fig. 19) the difference in reactivity of the cortical and medullary bone is manifest by the intensity and shade of colour they take on. The latter stains an intense purplish-red.

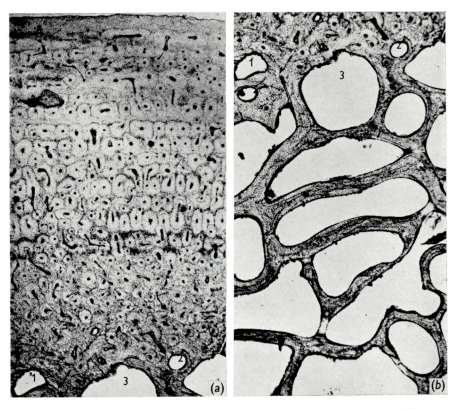

Fig. 20. Ground transverse section of long bone, magnification, × 12. (*a*) and (*b*) are photographs of adjoining areas. The former shows the compact bone of the cortex, the latter the cancellous bone of the outer medulla. The relationship of (*a*) to (*b*) can be seen from the positions of the spaces marked 1, 2 and 3.

In (*a*) the cortical vascular, soft tissue spaces or canals are small; in (*b*) they are altogether much larger and are surrounded by relatively thin trabeculae of lamellar bone of structure comparable to that shown under higher magnification in Figs. 16 & 17. The soft tissues which occupied the spaces in life were lost during the grinding of the section.

Examination of a transection of a tubular bone shaft shows that the density of its tissue varies from periphery to centre. The variable is the ratio of bone tissue proper to the soft tissue (blood vessels, fat, myeloid tissue, nerve) enclosed within it. This variation in density or porosity is responsible for the micro-architecture of the bone and may be demonstrated either in sections or by study of the cut surface of a bone.

Beneath the periosteum is an outer zone, very dense, consisting of circumferential lamellae. In sectional view, these look like stacks of lamellae piled on top of each other with long axes parallel to the bone

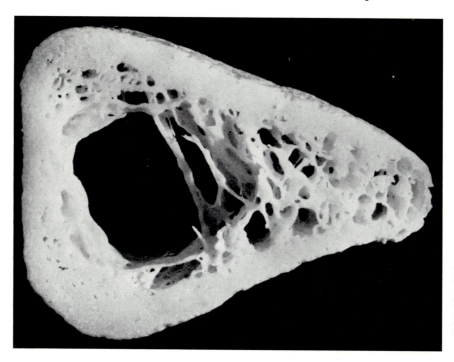

Fig. 21. Segment of human long bone which had been extracted with ether and treated with hydrogen peroxide. Magnification, ×4.50. The differences in porosity between the compact bone of the cortex and the cancellous bone of the medulla are apparent.

surface. In reality they are sheets wrapped around the bone circumference. Deeper, there are concentric lamellae arranged around Haversian canals, more or less parallel or oblique to the bone surface; then a substantial layer of Haversian systems seen cut more or less transversely (Fig. 20). Interstitial lamellae pack up and fill in everywhere between Haversian systems. This outer, dense zone comprises the *cortex*, and the bone tissue, as we have seen is dense or *compact*.

Travelling inwards, the spaces enclosed by the lamellae become greatly enlarged (Fig. 20), and would contain not only vessels but also fat and myeloid tissue. This more porous or *cancellous* bone tissue constitutes the *medulla*.

The construction of the bone tissue in the medulla is lamellar (Figs. 16 & 17) but as the vascular spaces are enormously expanded, the lamellae are seldom truly concentric. They usually have a radius of curvature greater than that of the vascular space itself. In any case the spaces are seldom round; they are often compressed ovals.

3

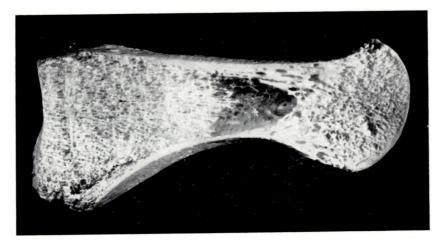

Fig. 22. Human metacarpal extracted in ether and treated with hydrogen peroxide, then cut longitudinally. Magnification, × 1.60. The relatively thin shell of compact cortical bone encloses a mass of cancellous medullary bone tissue.

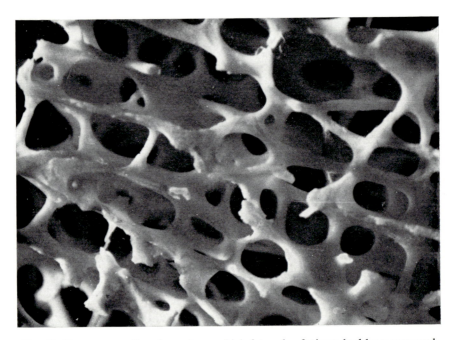

Fig. 23. Human cancellous bone from which fat and soft tissue had been removed. Magnification, × 15. The anastomosing sheets and bars of lamellar bone surround spaces which contained the vascular and soft tissue elements.

The differences in texture between compact and cancellous bone may be seen clearly at the cut surface of bones, as well as in sections. A clear picture is given by specimens treated beforehand to extract fat and to remove their soft tissue content; for example, by extracting with ether in a Soxhlet apparatus and then treating with hydrogen peroxide (Figs. 21, 22 & 23).

The relative numbers and cross-sectional diameters of the Haversian systems, and of the interstitial lamellae, have been said to vary from one bone of an individual to another. Evans & Bang (1966) maintain that fibular bone has relatively few, though large, osteones and plentiful interstitial lamellae. (This might well indicate a high rate of bone turnover.) Femoral bone on the other hand has many small osteones and sparse interstitial lamellae.

In summary, it can be said that lamellar bone is a material built up of unit layers. These contain fine collagen fibres which, in any unit, all run in approximately the same direction; but the axis differs by about 90° in adjacent units. The layers may be arranged more or less concentrically around a vascular canal; the lamellae grouped around a particular canal form a Haversian system. Other, interstitial, lamellae run in groups between the Haversian systems; as we shall see later, these are the last remains of previously-existing, complete systems which have been resorbed in the course of the normal turn-over of the matrix. Osteocyte lacunae are arranged in a characteristic way, concentrically disposed in concentric lamellae. Their canaliculi communicate freely, adopting in general a radial arrangement with respect to the central canal. Lamellar bone may be *compact* (small vascular spaces) or *cancellous* (large vascular and soft tissue cavities).

4

BASIC COMPONENTS OF
BONE MATRIX

In the previous chapter, brief mention was made of the constituents of the intercellular bone matrix, i.e. the collagenous framework, the protein–polysaccharide ground substance and the bone mineral. The ways in which these constituents are put together in the construction of bone matrix was explored chiefly from the histological point of view. Their biochemical and biophysical aspects were scarcely touched upon, and will now be referred to in more detail.

1. COLLAGEN

About 95 per cent of the bulk of bone tissue consists of the protein collagen in the form of white collagenous fibres; it is the collagen framework which confers shape upon a bone. With techniques currently available, the collagenous fibres of bone are indistinguishable morphologically from those of other collagen-rich tissues like skin or tendon.

Collagen is a fibrous protein which is made from aggregations of tropocollagen macromolecules. The tropocollagen macromolecule, the basic structural unit of collagen, is formed of three separate polypeptide chains which intertwine in the form of a left spiralling triple helix. The triple helix itself winds in a right-handed spiral around a common central axis (Ramachandran, 1963). This molecular structure gives rise to an X-ray diffraction pattern that is common to all collagens indicating the structural similarity of, for example, bone, tendon and skin collagens. The three polypeptide chains vary slightly in chemical composition, two being similar and called the alpha 1 chains, the other, the alpha 2 chain, being chemically somewhat different, containing a higher proportion of basic amino acids (molecular weights for each chain about 95000) (Piez *et al.* 1963).

Each of the alpha chains is composed of amino acids linked together by the elimination of water between the carboxy group of one amino acid and the amino group of its neighbour. The chemical composition of collagen is unique in that glycine forms one third of the amino acids and proline and hydroxyproline about a further quarter.

OH
|
CH——CH$_2$
| |
CH$_2$ CH—COOH
   NH
Hydroxyproline

COOH
/
CH$_2$
\
NH$_2$
Glycine

CH$_2$—CH$_2$
| |
CH$_2$ CH—COOH
   NH
Proline

There is a close similarity in amino acid composition of collagen from bone and skin in different species and different stages in maturity. A slight species difference occurs in the amount of valine and alanine, but, generally speaking, the differences are very small (Miller & Martin, 1968). Even in fossil collagens, the amino acid composition is very similar (Ho, 1967).

The amino acid residues which make up the alpha chains may be electrically neutral, e.g. alanine, valine; or may be charged positively, e.g. lysine, arginine; or negatively, e.g. aspartic acid, glutamic acid. The distribution of charged residues along the tropocollagen monomer is asymmetrical and the two ends of the molecule differ; in effect there is a 'head' and a 'tail' to the monomer (Cooper & Russell, 1969).

The tropocollagen molecule is about 2800 Å long, and 13.6 Å diameter; it has a molecular weight of about 290000. Under normal conditions, lateral aggregation of tropocollagen molecules, overlapping longitudinally by specific amounts, form the collagen fibrils visible by electron microscopy.

The characteristic morphological feature of such fibrils is a periodic banding whose repeat distance along the fibril is roughly 640 Å. If collagen is positively stained with uranium ions (positively charged) or phosphotungstate ions (negatively charged) further sub-periodic banding is revealed. The periodic and sub-periodic banding is a function of the aggregation of tropocollagen molecules. It is believed that, during aggregation, charged regions along the tropocollagen monomer become aligned, so yielding the transverse sub-periodic banding.

Schmitt *et al.* (1955) proposed that the major, 640 Å periodicity arose through an overlap of the tropocollagen molecule with respect to its nearest neighbours by about one quarter of its length. This was later modified so that linear arrays extended over five 640 Å periods; diagram 1*a* (Petruska & Hodge, 1964). This modified 'quarter-stagger' theory has been criticized by Grant *et al.* (1965) who, on the results of negative staining, proposed that the tropocollagen molecule is divided into nine regions longitudinally, namely, five 'bonding' and four essentially non-bonding zones (diagram 1*b*). The five bonding zones, though not identical, are sufficiently similar for an equal probability to exist for any bonding zone on one macromolecule to bind (by intermolecular electrostatic and hydrogen bonds) to any bonding zone on another macromolecule. This hypothesis allows for some morphological flexibility in the tropocollagen

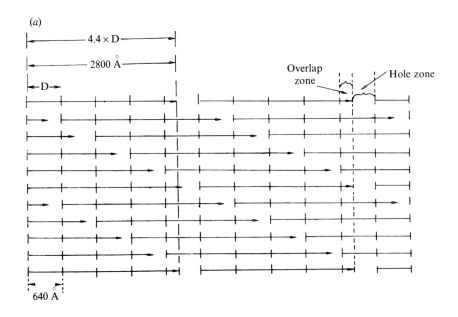

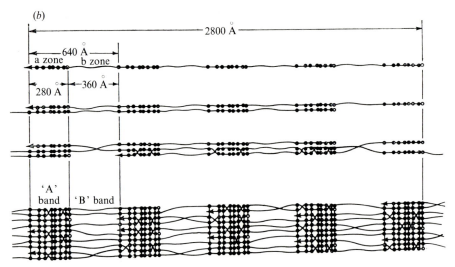

Diagram 1. The asymmetry of tropocollagen is represented by an arrow head at one end of the molecule.

(a) Illustrates the packing of 2800 Å tropocollagen macromolecules in linear array extending over five 640 Å periods (D). The hole zone is 0.6 D and the overlap zone is 0.4 D. (After Hodge *et al.* 1965.)

(b) Illustrates the formation of a 'native type' collagen fibril by random lateral aggregation of 2800 Å long tropocollagen macromolecules. The tropocollagen macromolecule is divided into five main bonding (a) zones and four main non-bonding (b) zones. The macromolecules can cross one another in both 'A' and 'B' bands, an indication of morphologic flexibility. (After Cox *et al.* 1967.)

(Reproduced with modifications by kind permission of the authors and publishers).

monomer and macromolecules may cross one another in the formation of fibrils.

The quarter-stagger model for tropocollagen aggregation has recently been further modified to overcome shortcomings in the earlier model, but structurally the 'random aggregation' model of Grant *et al.* seems more easily to fit the known properties of collagen. (For a review of the molecular structure of collagen see Cooper & Russell, 1969.)

Experiments designed to study the products obtained by dissolution of collagens in a variety of solutions have led to a great increase in the knowledge of its molecular aggregation. The major difference between bone and other collagens is found in the solubility of these materials in neutral salt and dilute acid solutions. The determination of the two different types of alpha chains was carried out by extracting skin collagen with cold 0.2 M NaCl solution at neutral pH, then separation by ion exchange chromatography. Up to 10 per cent of skin collagen could be extracted and separated in this way. By using stronger salt solutions followed by solutions of acids such as acetic and citric, more collagen could be extracted from tissue (Jackson & Bentley, 1960); but the collagen could always be re-precipitated, either by dialysis from salt solutions, or by careful neutralization, to yield fibrils showing, by electron microscopy, the same band pattern as the original material. When similar techniques were applied to relatively young bone such as three week old chick tibiae, less than 1 per cent of the total bone collagen was dissolved in 0.5 M acetic acid (Miller & Martin, 1968).

By treating the neutral salt-soluble collagens with reagents which disrupt hydrogen bonds, or by heating, the alpha chains can be separated, indicating that strong covalent forces do not bind them together. The acid-soluble collagens treated in this way may contain dimers of the alpha chains. If progressively more concentrated acids are used to solubilize collagen, a proportion of the alpha chains is left as trimers; that is, three alpha chains firmly bound together. These results indicate that the alpha chains, in collagens soluble only in relatively severe conditions, are bound together by strong covalent bonds. Thus, the less soluble collagens have a more stable chain configuration. Other links between the alpha chains are also thought to form as a process of ageing. Ester bonds have been reported which may occur directly between two chains or through an intermediate carbohydrate group.

In addition to this evidence for intramolecular covalent cross-links, it appears that covalent links bind adjacent macromolecules together in the less soluble collagens (Bornstein & Piez, 1964). It has been suggested by Veis & Anesey (1965) that the collagens insoluble in relatively severe conditions (neutral solutions at temperatures of or below 60 °C) are proteins which are highly cross-linked both intra- and intermolecularly.

By using 5–10 M solutions of lithium or guanidinium salts, further quantities of collagen can be dissolved from both skin and bone. However, whereas the solubilized skin collagen could be separated into alpha chains and dimers, the bone collagen tends to show the existence of aggregation states in the solution and also signs of degradation, that is, molecular weights of less than the expected approximately 95000 (Miller *et al.* 1967).

Glimcher & Katz (1965), from a comparative study on bone and tendon collagens, believe that the bone collagen macromolecules are bound by strong non-covalent intermolecular forces and differ in this respect from other collagens. They believe that the solubility of collagen macromolecules is a function of the strength and number of intermolecular bonds and forces, and that an increase in the number and strength of intermolecular bonds during collagen maturation causes it to be more insoluble in salt and acid solutions. The nature of the hypothetical non-covalent bond is not discussed in their work.

At present it is certain that bone collagen is similar to other collagens both in amino acid composition and in its aggregation to a coiled three chain structure, tropocollagen. Bone collagen is more resistant to dissolution in salt and acid solutions, which is probably a function of increased molecular binding, but whether this is due to a more rapid build-up of covalent bonds or whether some as yet unspecified non-covalent bonding occurs is not clear. Why, in normal circumstances, mineral deposition will take place in bone collagen and yet not in other collagens is equally obscure, but differences outlined above between bone and other collagens may have an important bearing on this phenomenon.

From the morphological point of view, as we have seen in Chapter 1, the collagen fibres of bone fall roughly into two classes, the coarse fibres of woven and the finer of lamellar bone. It is also possible to classify bone fibres in a slightly different way into two groups, depending on their origin. On one hand are fibres constructed in the normal course of events as part of the matrix of newly-forming bone, being synthesized, in the ordinary way, by the matrix-producing osteoblasts. On the other hand, and most often seen where osteogenesis proceeds rapidly, as in fracture repair, are bundles of collagen fibres which have obviously been laid down earlier, as 'connective tissue' fibres, and subsequently utilized and incorporated into the advancing edge of newly-forming bone (Fig. 33). Here it seems as if osteoblasts had availed themselves of acceptable material already to hand, whether amino acids or whole masses of prepared collagen fibrils; such bone seems to calcify normally. This is a convenient point to remark that other materials besides collagen may be incorporated in rapidly growing bone. Elastic fibres are not uncommon in fracture callus; under experimental conditions, cholesterol crystals scattered on a growing bone surface can be recognized later on, buried beneath the surface.

Since the substance of the bones is 'turned over' (removed and replaced) as a continuing, normal process, as much collagen must be broken down and its degradation products removed as is synthesized. As Eastoe (1968) points out, it is rather curious that for the animal kingdom, in which collagen is so widely distributed, such a complex supporting material has been evolved; in the plant kingdom, the main 'skeletal' substance, cellulose, is much simpler.

2. GROUND SUBSTANCE

The evidence that there is something else in bone matrix besides collagen and bone salt comes from both microscopical and chemical work. In bone sections stained by methods which pick out the collagenous fibres vividly or impregnate them with silver, an unstained background can often be detected between the fibres or fibre bundles with the light microscope. This, in a broad way, probably represents a negative image of the ground substance. After staining with the Hotchkiss PAS technique, the matrix gives a positive reaction, which varies according to the nature of the bone sample. Newly-formed bone, whether woven or lamellar, reacts more strongly than older bone (Fig. 19). The positive reaction, of course, proves the presence of mucopolysaccharide, i.e. protein–carbohydrate complex. So far, attempts to demonstrate the ground substance of bone at the electron microscope level have not succeeded.

The study of the protein–polysaccharides associated with connective tissues by means of electron microscopic methods is still more or less in its infancy. However, the idea seems to be growing that at least in certain situations, like the cornea, 'the ground substance is not a random structure filling the spaces between the fibres, but, rather, a well-defined, complex molecular chain system providing rigid bridges between the fibres and thereby being responsible for the maintenance of the regular array of collagen fibres' (Balazs, 1965). Smith and colleagues (Smith *et al.* 1967; Smith & Serafini-Fracassini, 1968; Smith & Frame, 1969) have worked with electron microscope techniques in this field and claim (Smith & Frame, 1969) to have vizualized fine filaments (about 40 Å wide and 2000 Å long) which, they think, are the protein cores of the protein–polysaccharide macromolecules. They give diagrams to explain how these filaments might 'lock on' and hold the associated collagen fibrils in place. Again, Meachim and his colleagues have recently demonstrated a meshwork of fine filamentous material in the intercellular matrix of both articular cartilage and nucleus pulposus of intervertebral disc. It seems quite likely that this represents protein–polysaccharide.

Results such as these are extremely interesting and show how molecular biology is overtaking tissues as well as cells. However, what goes for the cornea, the intervertebral disc or even cartilage may not necessarily hold for bone.

The organic matrix of bovine cortical bone consists principally of the fibrous protein collagen together with an amorphous protein–mucopolysaccharide complex and a proteinaceous material resistant to gelatinization (Table 1).

TABLE 1. *Composition of bovine cortical bone (wt %)*

Inorganic matter insoluble in hot water	69.66
Inorganic matter soluble in hot water	1.25
Collagen	18.64
Mucopolysaccharide–protein complex	0.24
Resistant protein material	1.02
Water (lost below 105 °C)	8.18
	98.99

(From Eastoe & Eastoe (1954), by permission of the authors and publishers.)

The protein–mucopolysaccharide matrix can be separated into a non-collagenous protein and a mixture of mucopolysaccharides, chondroitin-4-sulphate, chondroitin-6-sulphate, keratan sulphate and sialic acid. The identity of the resistant protein was not established by Eastoe & Eastoe (1954), and has still not been characterized.

The chemical structure of the disaccharides which form the basic repeating units of the polysaccharides is as follows.

Chondroitin sulphate is composed of D-glucuronic acid and N-acetyl galactosamine with a sulphate group attached to the carbon of C_4 or C_6 respectively of the galactose molecule. Keratan sulphate is composed of galactose and N-acetyl glucosamine, the latter bearing a sulphate group on C_6. The molecular weight of the protein–polysaccharide complex in bovine nasal cartilage has been estimated using several different techniques but with rather conflicting results. A low value of 2.4×10^5 was found by osmometry and a high of 4.6×10^6 by viscosimetry and light-scattering measurements (Campo, 1970).

A model for the structure of protein–polysaccharide matrix has been proposed (Matthews & Lozaityte, 1958) which consists of a central protein rod about 3700 Å long, to which several chains of chondroitin sulphate are linked. The protein core is disordered and appears to lack a secondary structure, that is, the protein molecules are probably not spirally linked as in collagen. This structure is thought to be dispersed in a random manner entangled amongst the collagen fibrils in the organic matrix.

The cartilage of the growing epiphyseal plate contains relatively large quantities of mucopolysaccharide (23 g/100 g acetone-rinsed tissue), but, towards the zone of mineralization, this amount decreases rapidly and where calcification is commencing in the primary spongiosa only 0.6 g per 100 g is found. In the secondary spongiosa this value is halved (Campo & Tourtellotte, 1967). Weatherell & Weidmann (1963) have shown that the newly-formed bone replacing endochondral calcified cartilage has a concentration of mucopolysaccharide little different from that of mature cortical and cancellous bone. Lindenbaum & Kuettner (1967) worked with calf scapula from which it is possible to 'harvest', for biochemical tests, ordinary cartilage at one end, calcifying cartilage and ossifying cartilage in the middle and newly-formed bone and older bone at the other end. Thus, they were able to compare the chemistry of the ground substance from these different sites. They found that whereas the ossifying cartilage contained 37 per cent of mucopolysaccharide by weight of organic matter, the figure fell to 4 per cent for new bone and 2 per cent for older bone. There were considerable differences in the amino acid constitution of the ground substance in these zones.

It has long been believed that changes take place in the ground substance where osteoid (the newest-formed organic bone matrix, not yet calcified) is to receive its quota of bone salt, or where bone matrix is to be resorbed, and it seems very likely that the amount of ground substance bears some kind of inverse relationship to the age of bone. Heller-Steinberg produced evidence of a change in the PAS reactivity of bone matrix resorbing under the influence of parathyroid hormone injections (1951), and there is a very striking difference in the reactivity of the medullary and cortical bone of the shaft of the femur in the laying hen. The medullary bone can be regarded as a sort of bank from which mineral may be withdrawn rapidly to provide calcium for the egg shell; conversely, it stores calcium when opportunity arises. The actual withdrawal and deposition are mediated respectively by resorption and deposition of bone. Medullary bone is thus extremely active, metabolically. It gives a very strong positive PAS reaction (Fig. 19), whereas the more inert shaft bone reacts very weakly.

In conclusion, it is worthwhile pointing out that much more is known about the ground substance of cartilage than of bone. For one thing, there is much more of the former and, for another, its role (in some cases at least) is much clearer. For example, it seems to be the chondroitin-sulphate-rich ground substance in the cartilage of rabbit's ears that keeps them upright; following the injection of papain into a young rabbit its ears flop, because the chondroitin sulphate leaves the matrix which shows loss of metachromasia. In a few days, the ears prick up again as the ground substance reverts to normal (McCluskey & Thomas, 1958).

The functional significance of the ground substance remains to be identified. It is conceivable that it plays a role in the calcification of bone; it may, as in the cornea, be involved in providing molecular bridges between collagen fibrils; and it may be the agency through which the hydration of the matrix is regulated. Bone matrix is not chalky dry in life, but detectably moist, and the likeliest candidate for holding and releasing water is the all-pervading ground substance. A comprehensive review on the protein–polysaccharides of cartilage and bone in health and disease has been published recently by Campo (1970).

3. LIPIDS

Recent work by Irving & Wuthier (1968) has shown the presence of small amounts of lipids in cancellous bone. Roughly 0.47 per cent of lipids was extracted from cancellous bone before decalcification and a further 0.14 per cent (both figures based on the demineralized weight of bone) from the demineralized sample. This is equivalent to 0.12 per cent lipid on the fully mineralized weight of cancellous bone. More than ten times this amount was found in hypertrophic and calcifying cartilage, indicating a decrease on bone formation akin to that of mucopolysaccharide.

4. BONE SALT

Bone is stony hard because it is impregnated with mineral, the bone salt, which is locked firmly in place in the organic supporting framework. In much the same way as for ground substance and collagen, the bone mineral, too, can be discussed from morphological and chemical standpoints.

The study of bone mineral morphologically is beset by technical problems, some of which were mentioned earlier (p. 10). There is first of all the problem of making sections at all and then of being sure that no mineral has been lost or displaced before the preparation reaches the microscope. These problems have to a large extent been overcome, but the possibility of artefactual loss of salt or of changes in its apparent distribution always remains a potential hazard.

The histological appearance in the light microscope of bone salt deposits is now really a matter of historical interest, though their actual concentration as investigated by microradiography is still very much in the news. The application of the methods perfected by Bloom & Bloom (1940) produces sections in which the bone salt is blackened by the von Kossa procedure. If, for example, a trabeculum of newly formed woven bone is examined after such preparative techniques, the more mature central area is seen to contain a homogenous blackened core, perhaps slightly granular in appearance. Here, mineralization has been completed; the full quota of bone salt has been laid down. The outermost edges of the newly deposited bone completely lack blackened material, as a time lag occurs between matrix formation and its mineralization. Between the mature, central area and the mineral-free, peripheral shell is the zone of commencing calcification where isolated aggregates of small blackened granules of varying size can be seen (Fig. 32). Owing, of course, to the limited resolution of the light microscope, the finer structure of these granules cannot be seen. Nevertheless many important facts about the calcification of bone matrix were obtained by these methods a decade or more before the electron microscope and its infinitely more fruitful techniques came to bear on this particular question. For example, facts were established about the time lag between the laying down of bone matrix and its calcification, in embryonic and newborn bones (Bloom & Bloom, 1940), in normal growing bones (McLean & Bloom, 1940) and in fracture repair (Urist & McLean, 1941). It was also shown that, in bone resorption following injections of parathyroid hormone, the organic matrix and the bone mineral seem to disappear together (McLean & Bloom, 1941).

The advent of methods for electron microscopy of bone naturally led to revolutionary discoveries about the bone salt. For instance, the tiniest aggregates of bone mineral visible with the light microscope could now be seen to consist of countless thousands of tiny needle- or rod-shaped crystals (Figs. 39 & 40). Naturally a host of other aggregates too small to be seen with the light microscope also come into view.

The bone salt crystal is thus of a size which falls within the resolution of the electron microscope; its average diameter seems to be about 30 Å and the length varies from a few hundred to a few thousand. The crystals seem to be in closely packed, shoal-like formations, parallel to each other (Fig. 40). The nature of the forces that hold the crystals in place, their precise sites of attachment to the organic matrix and the factors which initiate their deposition are best left to be discussed more fully in the section devoted to calcification mechanisms in Chapter 8.

A further technique much developed in recent years to study the bone mineral is, of course, microradiography, which has revealed that calcification is unequally distributed. The oldest bone matrix, that comprising the interstitial and outer lamellae of Haversian systems, turns out to be the most heavily calcified, whilst the less calcified, innermost layers of the systems of other newly-forming bone are markedly radiolucent (Fig. 1). Other methods for studying calcification, of course, include the use of the various tracers mentioned briefly in Chapter 2.

5. CHEMISTRY OF BONE SALT

Bone mineral is a complex chemical made from calcium, phosphate and hydroxyl ions, but which may also contain small amounts of cationic magnesium and strontium, replacing calcium, and bicarbonate and fluoride, replacing the hydroxyl anions. The crystals are very small. Whereas a crystal of a substance such as common salt or washing soda may grow to 10 millimetres or more in any dimension, the crystals of bone mineral rarely exceed 50 Å in thickness and are 400–2000 Å in length. The crystals may be needle-shaped, or plate-like in character. The size of the inorganic crystals of bone was first roughly determined by X-ray diffraction (Engström, 1960), but later, electron microscopy was able to provide pictures from which reasonably accurate measurements could be taken.

The arrangement of the atoms within these tiny crystals has been derived by X-ray diffraction (Posner *et al.* 1958) and it was shown that the structure of bone mineral resembled the geological apatites. Apatites form a class of compounds which contain (*a*) divalent cations, (*b*) tetrahedral trivalent anions and (*c*) monovalent anions. Bone salt in particular contains a calcium cation and orthophosphate and hydroxyl anions. The exact chemical composition is not known because of factors which will be mentioned later, but the basic formula is represented as Ca_{10} $(PO_4)_6$ $(OH)_2$, calcium hydroxyapatite.

In bone apatite, calcium may be replaced in part by other divalent ions such as lead, strontium, barium and radium, and the monovalent hydroxyl may be replaced by fluoride, in which cases the ionic distances will alter slightly, but the apatite character remains.

The crystal lattice structure of calcium hydroxyapatite in a direction parallel to the long axis of the crystal is formed of hexagons of calcium ions surrounded by hexagons of phosphate ions (Kay *et al.* 1964). These hexagonal arrays are built one on another to form the long axis of the crystal. The four remaining calcium ions (of the chemical formula) are arranged between the hexagons in the third dimension and the two hydroxyl groups lie in the centre of the hexagon along an axis also in the third dimension.

It is usual when examining the internal structure of crystals to use the term 'unit cell'. The unit cell is the smallest arrangement of the crystal ions which can be found repeated in the same arrangement and ratio throughout the whole crystal. This is why the bone mineral formula is expressed as Ca_{10} $(PO_4)_6$ $(OH)_2$ and not Ca_5 $(PO_4)_3$ OH.

In the unit cell of an idealized calcium hydroxyapatite, the *xy* plane (*ab* axes) lies inside imaginary lines joining the hydroxyl groups and forming a parallelogram with sides of 9.43 Å. The unit cell extends 6.88 Å in the direction of the *c* axis (Kay *et al.* 1964).

In crystals such as those of common salt, the size is such that many thousands of regularly repeating units or unit cells are present inside the crystal boundaries and any surface defect has a negligible effect on the X-ray diffraction pattern. In bone hydroxyapatite, however, the unit cell is large in comparison with the cross sectional dimensions of the crystal (10–30 Å in the *a, b* axes or 1 to 3 unit cells either way) and consequently a defect due to ions which are not normally a part of the hydroxyapatite lattice or due to a poor crystalline structure would materially alter the expected X-ray diffraction pattern. In fact, not only is the X-ray pattern of bone mineral poorly defined, but chemical analyses of bone and tooth mineral as well as synthetic apatites differ in composition from that expected of $Ca_{10} (PO_4)_6 (OH)_2$ (Brown, 1966).

The theoretical ratio of Ca/P in the above formula is 1.667 and if the crystals were large in relation to the unit cell this ratio would be approximately correct for the hydroxyapatite crystal. As previously pointed out, the crystals are very small, so that the number of ions at the crystal surface which are not shared form a relatively large proportion of the total ions in the crystal. This significantly alters the expected Ca/P ratio which then depends largely on the final crystal size. Experiments have shown that the Ca/P ratio increases with age (Pellegrino & Blitz, 1968) indicating that bone crystals gradually increase in overall size with skeletal age.

It is generally accepted that bone mineral has basically an apatite structure and many proposals have been advanced to account for the deviations from the theoretical formula (Brown, 1966). Apart from non-stoichiometry with which we have just dealt there have been suggestions that adsorption or occlusion of phosphate ions on the surface of the crystals could occur or that other cations (Na^+, K^+) could replace calcium ions, reducing the Ca/P ratio.

Other, likelier suggestions, are that the crystal itself might be physically different. For instance, phases other than hydroxyapatite might be present (Herman & Dallemagne, 1961); also, it has been shown that some tightly-bound water is present in biological apatites. The hydroxyapatite formula does not contain water. Winand & Dallemange (1962) proposed a formula for a calcium-deficient apatite with a Ca/P ratio of 1.5 in which an overall neutrality was maintained by hydrogen bonding. The compound forms pyrophosphate on heating, is considered to contain water molecules (by infrared spectroscopy) and has a Ca/P ratio similar to many biological apatites. Whilst calcium-deficient apatites of this nature are not now believed to form any significant part of mature bone, Termine (1966) has suggested that they may be formed as intermediates during crystal maturation.

Brown (1966) has suggested that octa-calcium phosphate is the initial crystalline phase precipitated in bone; by gradual hydrolysis and phase transformation hydroxyapatite is finally formed. The Ca/P ratio is only 1.33 in octa-calcium phosphate. The system proposed by Brown accounts for the non-stoichiometry of bone salt and for the variability of the Ca/P ratio which Pellegrino & Blitz (1968) have shown increases with age. It also allows for the formation of pyrophosphate (from octa-calcium phosphate) on heating.

It is fair to say that Brown (1966) himself has pointed out that it is very difficult to distinguish structurally between octa-calcium phosphate (OCP), tetra-calcium phosphate and calcium hydroxyapatite (CHA) because of their relatively similar X-ray powder diffraction patterns. In bone, of course, where OCP and CHA would be intermingled, the powder diffraction pattern is not able to show unequivocally the presence or absence of OCP. Eanes and his associates (Eanes & Posner, 1965; Eanes *et al.* 1965, 1966) have recently indicated that in the precipitation *in vitro* of calcium–phosphate mixtures from alkaline solutions the first solid phase is essentially non-crystalline. A poorly crystalline bone apatite is formed when this amorphous compound is allowed to stand in contact with the alkaline fluid.

These workers have also shown that both bone and cartilage contain amorphous as well as crystalline calcium phosphates and the relative proportion of amorphous calcium phosphate decreases with age whilst that of the poorly crystalline hydroxyapatite increases.

By electron microscopy, early deposits of bone mineral always appear to contain dense needle-like crystals together with a less dense material (Fig. 39 inset). If amorphous calcium phosphate is the initial phase of mineral deposition, such observations imply that there must be a rapid conversion of part of the amorphous solid to a crystalline form, or that initial deposits of amorphous mineral, which are expected to be perhaps 5–10 Å diameter, are too small to be detected in thin sections by electron microscopy. The size of deposits containing even the smallest recognizable crystals are large in comparison with the theoretically smallest ion cluster able to separate from solution (Katz, 1969). Höhling (1969) states that thin rods develop from dot-like 'nuclei' during mineralization of bone, though such findings have not been observed by the present authors (Fig. 39 inset). Later, plate-like crystals with a more highly developed crystallinity form from the thin rods. The plate-like crystals appear to be a carbonate-rich octa-calcium phosphate rather than hydroxyapatite. Bocciarelli (1970) also claims that the crystals in bone are morphologically a carbonate-rich octa-calcium phosphate, but her results are based on an analysis of mature bone crystals. By using specimen tilting she was able to show that a proportion of crystals which were rod-like in one plane were clearly plate-like in character when the angle was altered through 40°.

These results lend support to the proposal by Brown (1966) that octa-calcium phosphate has a role to play in biological calcification, but morphological evidence shows that an amorphous solid and a needle-shaped crystallite, probably apatitic in character, are the first recognizable mineral structures in bone calcification.

It has been shown recently (Lörcher & Newesely, 1969) that calcite (calcium carbonate) probably occurs as well as apatite in the medullary bone of the laying hen. Calcite of course is of widespread distribution in the invertebrates but its role in the vertebrate remains to be established. It is known that medullary bone acts as a sort of short-term deposit bank to accept and pay out calcium in response to the demands of the shell gland, and so a possible inference is that calcite can be more rapidly mobilized from (and taken into) bone than apatite.

During the last ten years or so interest has been growing in a completely new idea about an intrinsic control mechanism within bones, based on the generation of piezoelectric changes emanating from changes in pressures on, or movements of its crystals. This, apparently, can work in two ways. When bones are deformed experimentally, detectable electric currents are generated. Conversely, the application of an electric charge is said to produce alterations in the architecture of the bone; for example, Bassett *et al.* (1964), one of the pioneers and chief contemporary enthusiasts in this field, implanted insulated batteries in dogs' femurs and claimed that electric currents induced massive osteogenesis. Lavine *et al.* (1968) also claim that D.C. current flow enhances new bone formation. However, it now seems that not only bone salt crystals but collagen (Lavine *et al.* 1968) and indeed many other biological materials, even wood, are piezoelectrogenic. A review article has been published lately by Bassett (1968).

5

EVOLUTION OF BONE

The fossil record shows that the occurrence of bone-like tissue dates back to the first vertebrates, the ostracoderms. These were jawless, armoured fish which lived a benthonic existence in the seas from the Ordovician (400 million years ago) to Devonian (250 million years ago) eras. Their skin contained dermal plates of various sorts and sizes, constructed of hard mineralized tissue (Fig. 24). As described in more detail below, this tissue was very similar to or even identical with modern bone in some species, whilst in others there were fundamental differences. The real function of the dermal plates and the evolutionary 'pressures' which may have been involved in their appearance are not yet settled (Berrill, 1955). According to some authors, the plates acted as a protective armour, either against predators, or (Smith, 1953), against unfavourable osmotic changes in the surrounding watery environment. It has also been suggested that the acquisition of relatively heavy bone masses may have helped their owners by adding displacement and so reducing the energy required to live a life submerged near the sea- or lake-bed (Schaeffer, 1961), from the mud of which the ostracoderms extracted their food.

It is still uncertain whether the first ostracoderms were sea- or fresh-water dwellers, though the former at present seems the more widely held view. If correct, this would make it improbable that dermal armour arose primarily as a store of calcium, since the latter would have been abundantly available at all times in sea-water (Denison, 1963). However, there might have been considerable advantages in having a reservoir of phosphate in the bone to iron out seasonal environmental variations in concentration. Such a source of phosphate would, of course, have been tremendously valuable when the land was being invaded later on by descendants of the ostracoderms (Pautard, 1961).

Whatever the real value of the dermal plates and the whys and wherefores of their evolution, their histological structure is extremely interesting. Two main methods are now available for preparing sections of fossil bone for microscopical examination at the light and electron microscope levels. First, as has been done for years, a slice of fossil-bearing mineral or the fossil itself may be taken and ground down by hand (for technical details, see Enlow & Brown, 1956, 1957, 1958; Ørvig, 1965). Such sections are

(a)

(b)

Fig. 24. Pieces of fossilized dermal armour, the outer surface being seen from above. Magnification, × 10.

(*a*) From *Tessaraspis*. The outer surface is adorned with raised, mamilliform 'denticles'.

(*b*) From *Pteraspis*. The outer surface is raised into an ornate pattern of ridges which, histologically, are rather similar to teeth.

naturally rather thick which precludes the use of higher magnifications; the presence of extraneous mineral in the sections may be distracting or even misleading and the technique is difficult and time-consuming. However, it has provided most of the information available to palaeomicroscopists.

The second method stems from the surprising discovery that much of the original organic material may actually endure in fossilized tissue. This was first demonstrated qualitatively, by paper chromatography of fossil residues (Abelson, 1956) which revealed the persistence of some of the collagen-constituent amino acids. Quantitative methods were applied soon after to residues (Ho, 1965; Armstrong & Tarlo, 1966; Tarlo, 1967). The upshot of these biochemical studies is that many of the amino acids found in collagen may persist in measurable quantities for millions of years in fossils, though their proportions may differ somewhat from those of modern bone collagen. For one thing, hydroxyproline, a hallmark of collagen, seems to be absent. Possible explanations are that different amino acids decay at different rates, and that there may be a differential

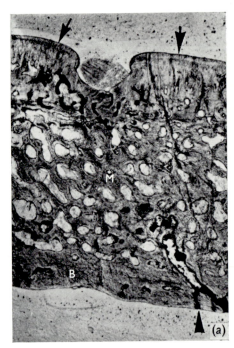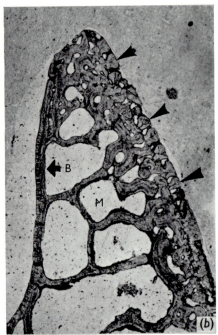

Fig. 25. Ground sections of fossil dermal armour plates; magnification, ×25.

(a) From *Tessaraspis*. Portions of two of the 'denticles' seen in Fig. 24a are present at the top (upper vertical arrows). The basal layer (B) is quite thick; in one place (lower vertical arrow) it is penetrated by a substantial canal which presumably conveyed blood vessels to the interior of the plate. The middle layer (M) contains numerous cavities, presumably containing blood vessels and soft tissues in life.

(b) From *Pteraspis*. The surface ridges seen in Fig. 24b are cut transversely (oblique arrows). The basal layer (B) consists of a few parallel laminae of aspidin whilst the middle, cancellous layer (M) comprises trabeculae of aspidin arranged around quite large spaces.

diffusion of amino acids out of fossil to nearby rock (Armstrong & Tarlo, 1966; Tarlo, 1967). However, the biochemical analysis of fossil residues holds great promise for the study of bone evolution at the molecular level.

Returning to the microscopy of fossil bone, as shown by Moss (1961a) and by Tarlo & Mercer (1961, 1966) the mineral constituents can be removed from fossils by the methods employed routinely for decalcifying modern bone; a framework of organic material survives which can then be embedded, as usual, and thin sections cut in the ordinary way. The fact that an organic framework can persist after decalcification has certainly opened up new possibilities in the microscopic study of fossil

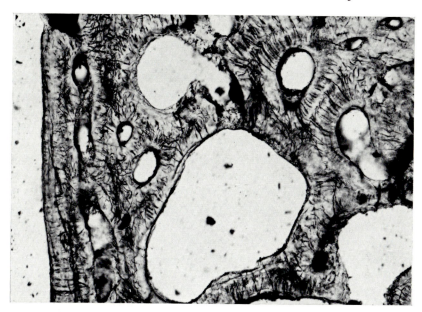

Fig. 26. Higher magnification view (× 120) of part of middle and basal layers of dermal plate of *Pteraspis*. The myriad spidery black lines within the trabeculae are 'aspidinocytes'.

bone, and has enabled work to be carried down to the electron microscope level (Isaacs *et al.* 1963; Doberenz, 1967; Wyckoff *et al.* 1963; Wyckoff & Doberenz, 1965). However, as discussed again in detail below, there are some problems in the interpretation of appearances.

The microscopic anatomy of the dermal plates is rather similar in the Osteostraci and the Heterostraci, the two ostracoderm families. Three well-defined layers are present (Fig. 25). The innermost or basal layer consists of stacks of parallel lamellae closely applied to each other. The middle layer, very much wider, consists of cancellous or spongy material; here the calcified tissue encloses spaces, much as in the epiphyses of a modern mammalian long bone. No trace remains of whatever soft tissues filled the spaces (which are probably interconnected) though they are generally believed to have been vascular. The outermost layer, elevated into ridges or tooth-like elevations of various heights and shapes and sizes, is made up of a dentine-like tissue and covered externally by a shiny material resembling enamel. These superficial dermal ridges and denticles were often beautifully and intricately ornamented.

Although the distribution of the mineralized tissue is rather similar in the dermal plates of the Osteostraci and the Heterostraci, there is a fundamental difference in the tissue itself. In the case of the former, it is bone,

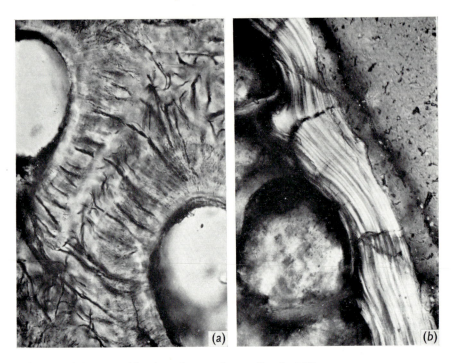

Fig. 27. Higher magnification of part of preceding (× 240).

(*a*) Parts of two vascular channels are surrounded by aspidin, through which 'aspidinocytes' run.

(*b*) Ground section of basal layer of fossil dermal plate from *Pteraspis rostrata*, photographed in polarized light (crossed nicols). The individual lamellae of aspidin are seen as parallel laminae. Aspidinocytes are not visible. Compare Figs. 16 & 17; the basic pattern of modern human lamellar bone is closely similar apart, of course, from the presence of osteocytes.

possessing the modern attributes of osteocytes in lacunae, whose processes, within canaliculi, course through the matrix. But in the Heterostraci there are no cells; this acellular matrix is known as aspidin. In the middle dermal plate layer, the aspidin is present in the form of concentric lamellae, quite reminiscent of the bone lamellae of the osteones of modern bone; such areas are known as aspidones. They often show clear evidence of 'model-ling'; that is, histological signs that both deposition (via 'aspidinoblasts') and resorption (via 'aspidinoclasts') must have taken place though, of course, no visible remains of cells persist in the fossils. Canaliculi radiate out from the innermost aspidinal ('endaspidinal') surface, somewhat reminiscent of the 'tubules in dentine (Figs. 26 & 27a). It is supposed that, just as dentine tubules carry odontoblast processes, so aspidin tubules conveyed those of aspidinoblasts. This suggests very strongly that aspidinoblasts formed the aspidin and, just as odontoblasts retreat in advance of the dentine they have formed, so with aspidinoblasts and aspidin.

The interpretation of this extraordinary tissue difference between osteostracan and heterostracan 'bone' is difficult and contentious. According to one school of thought (Tarlo, 1963, 1964; Halstead, 1969 a, b) aspidin should be regarded as more primitive than bone, for two principal reasons. It is claimed, first, that aspidin occurs earlier in geological time than true bone, which suggests that the latter arose through evolution from the former. Secondly, it is said, appearances prove that over a period of time of about 150 million years, extending from the Ordovician to Upper Devonian periods, aspidin gradually altered and acquired matrix cells, the aspidinocytes, which are regarded as the primitive precursors of the osteocyte. Aspidinocytes, as illustrated by Tarlo (1963, 1964) are certainly quite unlike osteocytes, being spindle-like or columnar in shape and with few or no processes (as in Figs. 26, 27 a & 28 a). According to this view, then, the heterostracons produced aspidin, which was itself the ancestor of bone.

A different viewpoint has been expressed by Ørvig (1965) who regards aspidin as derived from, rather than the source of cellular bone. He maintains that the same Ordovician strata yield both aspidin and bone. This of course, reduces the likelihood of the latter being the ancestral source of the former. In addition, Ørvig questions the real identity of the 'aspidinocytes' and suggests that, far from being lacunae occupied by cells, they represent coarser collagen fibres (or their remains) in the central areas of the aspidin trabeculae.

This seems at first thought to be the kind of question which could be pursued profitably by work with sections of decalcified dermal plate material, since, as was explained earlier, much thinner preparations can be had this way and the assistance of histological stains can be employed. Unfortunately, however, the interpretation of histological appearances in such sections can be very far from straightforward. For example, although 'aspidinocytes' were a prominent feature of ground sections made from the dermal plate illustrated in Fig. 24 a, they were not seen in sections of the same material after it had been decalcified (Fig. 28 b). The organic material left after decalcification was exceedingly tenuous and would not stain at all with dyes normally taken up by bone. Nothing was encountered to match the appearance of the 'aspidinocytes' seen in Fig. 28 a, though coccus-like bodies of equivocal nature were present. It is practically impossible to determine the nature of these bodies on present evidence. There seem to be three possibilities. On the one hand, they might indeed represent the sites of coarse collagen bundles, comparable to the Sharpey fibres of modern bone. On the other hand, they might represent tubules which, in life, had been occupied by 'aspidinocytes' or their processes; although

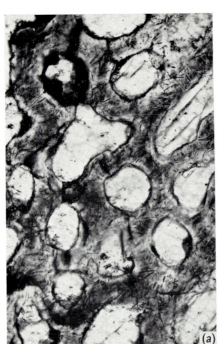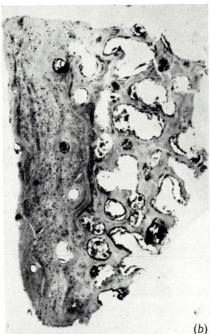

(a)   (b)

Fig. 28. (*a*) Ground section from *Tessaraspis*. Magnification, × 100. A part of the middle layer is shown. 'Aspidinocytes' are present.

(*b*) Magnification, × 50. Thin (4 μ) section from decalcified fossil aspidin (*Tessaraspis*). The specimen was partly decalcified in formic acid, then embedded in agar to minimize crumbling. Demineralization in formic acid was recommenced and, when completed, the agar block containing the sample was infiltrated in ester wax. Sections cut easily; haematoxylin and eosin.

The basal layer and part of the middle layer are present (see Figs. 25*a* & 28*a*). In the former, especially, are numerous small darkly-stained rounded bodies of uncertain nature.

there are far more of them, set much more closely together, than the osteocytes of modern bone. Finally, they might even represent the remains or effects of bacteria, yeasts or fungi growing in the material before it fossilized. This is known to have happened (Moodie, 1926, 1928). A further point is relevant to the ancestral connections of aspidin and bone. As was first shown over 100 years ago by Kölliker (1859), some of the higher orders of teleost fish have completely acellular bone tissue. He stresses the fact that in the teleosts the acellular bone has almost certainly evolved from cellular, which can be taken to indicate that the latter is more 'primitive' than the former. Before leaving this aspect of the ancestry of bone tissue, it should be pointed out that our knowledge is of course con-

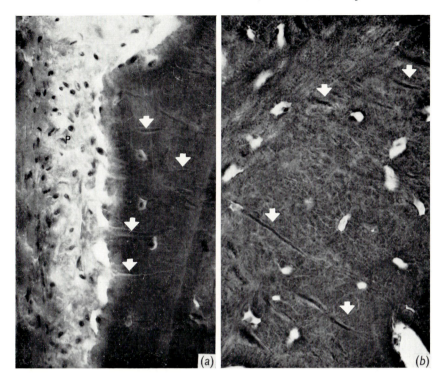

Fig. 29. Newborn human tibia. Decalcified, transverse section. Haematoxylin and eosin. Magnification, ×240.

(*a*) The periosteum forms a pale-stained layer (P) which has shrunk away from the bone edge in places. Prolongations of the collagen fibres of the periosteum can be seen to penetrate quite deeply into the subjacent bone (arrows).

(*b*) Deeply within the bone, similar fibres of the Sharpey type, are still easily visible (arrows).

fined to the limits of the fossil record as it is known at present. Something earlier may well turn up to revolutionize present views.

The ontogenesis of a bone-like hard tissue must have taken place far back in evolution, long before the present earliest fossil record. In the complete absence of any objective evidence, one has to fall back upon speculation as to how bone first made its appearance. It seems likely that the first steps must have been of a macromolecular nature, consisting of the development by the organism of the means of synthesizing the organic matrix and, thereafter, evolving the enzymes and so forth needed for its calcification. These two events, whether they took place stepwise or simultaneously, could have been followed by the morphological refinement of the hard tissue towards the microscopical pattern we recognize as characteristic for bone.

Collagen itself must be of exceedingly ancient ancestry. Naturally the fossil record is silent on this question, since collagen apparently survives only within the framework of fossilized bone or tooth. Comparative phylogeny, however, signposts a widespread distribution for this fibrous protein commencing with the porifera and coelenterata (Mathews, 1967; Chapman, 1966; Gross & Piez, 1960) and insects (Smith, 1968). It has indeed been suggested but, of course, not proved (Willmer, 1965, 1970), that collagen arose initially in an ancestral unicellular organism as a cell-surface regulator of metabolic traffic. During subsequent phylogenesis, the collagenogenetic trick was retained and 'diversified' towards a supportive role.

There seem to be no strong clues as to when the mucopolysaccharide which accompanies the collagen of bone first appeared. This event may have been involved in the beginnings of calcification. In the modern vertebrate, the tissues which normally calcify have a bulky collagenous framework which of course confers shape and size; but not all collagenous frameworks calcify. They only do so on acquiring an ill-defined quality of 'calcifiability'. The factors involved in the realization of this property are still uncertain, but it is possible that mucopolysaccharide may be specifically involved; this question is dealt with in more detail in Chapter 8. Obviously, the accession of calcifiability was a key event in the evolution of hard tissues. It may well have occurred soon after collagen appeared. Pautard (1961) described a unicellular organism containing what, in the electron microscope, constitutes the essential components of bone: clusters of hydroxyapatite crystals, scarcely distinguishable in appearance from those of modern mammalian calcified cartilage, in association with organic fibrils. Having made its appearance in the animal kingdom as a bodily tissue, the morphological pattern of bone has evolved in widely different ways in different species. It is interesting that the comparative aspects of bone structure made a very strong appeal to the early histologists (Kölliker, 1859; Quekett, 1846).

The components of modern bone tissue are laid out, at one end of the scale, as simple, closely-stacked lamellae without any cells (as in some teleosts) through homogenous bars, populated by osteocytes but completely lacking internal vascularization, as for example, in bats and insectivores, to the more complex arrangement characteristic of the higher mammals (Moss, 1968; Crawford, 1940). However, the dangers of reading too much into these differences as reflecting evolution of bone as a tissue have been well expressed by Enlow & Brown (1958), as follows:

The impression that the bone of all lower vertebrates is simple and primitive in structure, and that the developing bone of mammals recapitulates such phylogeny, is not justified. Neither can the bone tissues of mammals be regarded as a culmination of skeletal tissue evolution. Although many mammals exhibit well-developed, dense

Haversian bone, Haversian tissues are also found in a number of amphibians, reptiles, and birds. While several lower vertebrate groups possess simple, even non-vascular bone, a number of mammals also display simple, non-Haversian and non-vascular bone tissues.

Early mammal-like reptiles, the pelycosaurs, did not have typical mammalian bone tissue. Rather, their bone tissue structure follows the laminated, endosteal Haversian pattern common to labyrinthodonts and other early reptilian groups. Therapsids possess either a dense Haversian or plexiform bone tissue. The bones of living, morphologically primitive groups of mammals, including the monotremes, marsupials, and insectivores, are not composed of dense Haversian tissues. Their bone tissues, rather, are of a generalized, primary vascular design. Monotremes and marsupials have elaborate, primary vascular tissues, but Haversian replacement is usually absent or undifferentiated. The longitudinal, primary vascular pattern of many primates, on the other hand, commonly becomes replaced by extensive, dense Haversian tissues.

SECTION 2

BONE DEPOSITION

6

HISTOLOGY OF OSTEOGENESIS

The main histological events in osteogenesis are reflections of the activities of the osteoblast. The functions of this cell will be gone into more thoroughly in the next chapter when the relevant experimental evidence will need to be reviewed; there follows now a brief description of the micro-anatomical features of bone formation.

Developing embryonic bones or fracture callus provide good specimens for studying the histology of osteogenesis; the embryonic mandible is particularly useful. As illustrated in Figs. 6 and 7, it consists of anastomosing trabeculae of woven bone. As the jaw grows, extensive changes take place in the trabeculae. For instance, the bone around the developing tooth germs must be removed to make room for their expansion; and as the mandible itself expands in three dimensions, fresh trabeculae will be formed and others will become longer.

At the extreme growing tips or advancing edges of trabeculae, where osteogenesis is proceeding apace, osteoblasts can readily be distinguished by their characteristic appearance from neighbouring mesenchyme, endothelial or blood cells. They are plump cells with basophil cytoplasm arranged, generally, in a palisade-like manner (Figs. 6, 30 & 31); they secrete the bone matrix which, when first deposited, is known as 'osteoid' or preosseous matrix. This material differs from mature bone matrix in that it is uncalcified (Fig. 32).

As time goes by and the osteoblasts continue to secrete, more osteoid is laid down upon the original material; consequently, trabeculae thicken, and, if osteoid is added to their tips, also lengthen. Further, the older, deeper, original osteoid matures and becomes calcified, although if osteogenesis continues trabeculae will still possess osteoid rims. In decalcified sections, the older and calcified matrix stains darkly (Fig. 30a). In undecalcified sections, where bone salt has been blackened by a silver method (Fig. 32), calcified areas are seen to be 'buried' within uncalcified preosseous matrix.

A significant event next occurs in the older (and so thicker) parts of the trabeculae; here and there osteoblasts on the surface become trapped by the advancing edge of the bone and surrounded, at first partly, and then entirely, by the matrix, to become osteocytes (Figs. 30b & 32a). They

[63]

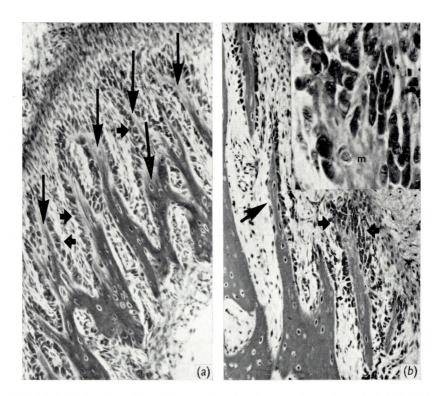

(a)

(b)

Fig. 30. (*a*) Advancing tips of bone trabeculae from a site such as that indicated on Fig. 6. Osteoblasts, arranged as palisades (e.g. horizontal arrows), have secreted bone matrix. The most recently deposited matrix of uncalcified osteoid stains palely (vertical arrows); the older matrix beneath, which was calcified, more darkly. Towards the bottom the trabeculae widen out considerably and anastomose. In these areas comprising the oldest bone some osteoblasts have become surrounded completely by matrix to become osteocytes. Magnification, × 120.

(*b*) Magnification, × 120. Active osteogenesis, with plump, polyhedral osteoblasts (horizontal arrows) and matrix between. The inset at top right illustrates a part of this area at higher magnification (× 480) and shows the osteoblasts and newly formed matrix (m) more clearly. Nearby (oblique arrow), a thin trabeculum has active osteoblasts along its right-hand edge, but, along the left, the cells are much flatter and relatively inactive. It seems likely that the osteocytes included within this trabeculum have been enclosed within the matrix as it spread to the right.

On the edge of the trabeculum forming the left-hand margin of the field the osteoblasts are very much attenuated and bone deposition was certainly at a standstill there when the specimen was taken.

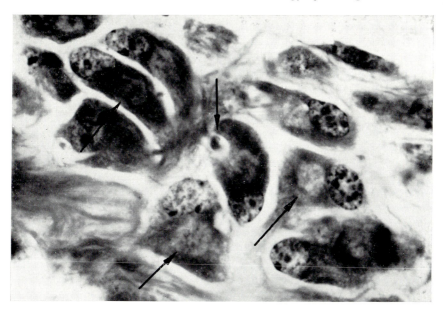

Fig. 31. Decalcified embryonic bone. Masson stain. Magnification, × 820. The field illustrates an early stage in osteogenesis. There is a group of about half a dozen active osteoblasts which show the typical cytoplasmic basophilia and the pale juxtanuclear area (e.g. oblique arrows). Nuclei tend to be positioned at one pole of the cell.

Between and around the osteoblasts is recently-formed bone, the 'preosseous matrix'. It appears as a filmy amorphous or slightly fibrillar material. At one point (vertical arrow) the cytoplasm of an osteoblast is indented by a minute trabeculum of this material; compare with Fig. 36.

occupy lacunae in the bone, and they have processes which traverse the matrix in canaliculi.

Meantime, two morphogenetic influences commence to influence the shape and size of the developing trabeculae. The first operates to control the variation in the output of bone by osteoblasts which may differ widely at different points and at different times. Osteoblasts which are actively osteogenetic appear as tall, plump, polyhedral cells (Figs. 30*b* & 31). Other osteoblasts, said to be resting, are flattened against the bone edge (Fig. 30*b*). In all probability, resting osteoblasts can be galvanized into activity again. As shown in Fig. 30, the shape of developing trabeculae must be determined, in part at least, by a differential in the bone output of different osteoblasts.

The second influence is exercised by the activities of a race of cells distinct from the osteoblasts, namely the osteoclasts, whose function is to erode ('resorb') the bone matrix. The osteoclasts are giant multinucleated cells; they are described more fully in Chapter 12. The osteoclasts remove

HBO

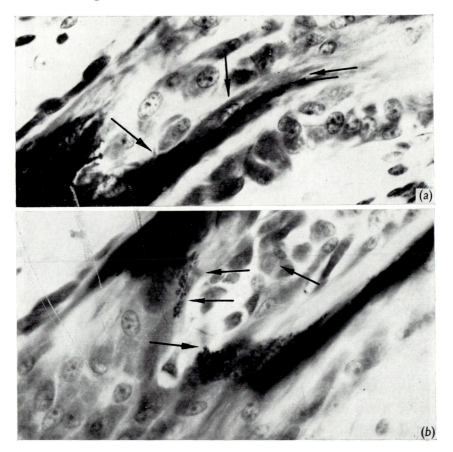

Fig. 32. Section of undecalcified cat embryo bone. Fixed in neutralized formol–saline, dehydrated in neutralized alcohols. Embedded in celloidin. Sections floated on neutralized water. Stained with haemalum and von Kossa silver nitrate method; bone salt appears black. Magnification, × 750.

(*a*) The oblique arrow points at a small trabeculum of newly-formed preosseous matrix. Near its growing tip (horizontal arrows), small islands of bone mineral can be identified as black areas. They are embedded in uncalcified matrix. Osteoblasts have become detached from the lower edge of the trabeculum. An osteoblast has become surrounded by newly-formed matrix (vertical arrow), and is now an osteocyte.

(*b*) The horizontal arrows point at blackened islands of bone mineral within uncalcified preosseous matrix. Osteoblasts are present; in one of them the pale juxtanuclear vacuole is visible (oblique arrow).

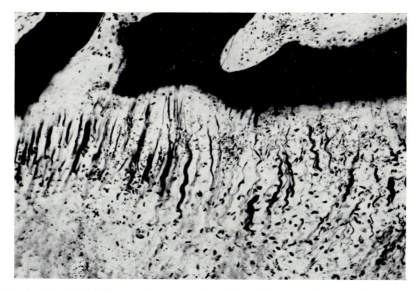

Fig. 33. Decalcified human fracture callus. Silver impregnation technique. Wavy collagen fibre bundles continue upwards into the advancing edge of a woven bone trabeculum. Elastic fibres are often seen to be similarly incorporated into bone. Magnification, × 120.

bone, whilst the osteoblasts deposit it. The interplay of these two opposite mechanisms is known as bone modelling. A good example is illustrated in Fig. 55. As the growing tooth germs expand the bone nearby is removed by osteoclasts, whilst osteoblasts further away lay down fresh matrix.

The foregoing account of the histological features of osteogenesis applies to the formation of bone in the embryo, and in postfoetal osteogenesis in general. However, where bone formation first occurs in fracture callus (and, occasionally, elsewhere), an additional mechanism is involved; existing collagen and even elastic fibres are utilized by osteoblasts and incorporated into the rapidly advancing edge of newly deposited bone matrix (Fig. 33). In some ways, this is reminiscent of the pattern of events which occurs in the ossification of turkey leg tendon (Johnson, 1960).

7

THE OSTEOBLAST

1. MORPHOLOGICAL FEATURES

The existence of a race of cells intimately associated with newly formed bone was first noted and illustrated by Tomes & de Morgan in 1853. The characteristics of these cells were described more fully by Gegenbaur, who gave them the title of osteoblast.

As mentioned in the last chapter, osteoblasts have a characteristic appearance. Seen with the light microscope, in fixed and stained preparations, these plump-looking, polyhedral cells have two cytoplasmic attributes (Fig. 31). Firstly, they possess a pronounced cytoplasmic basophilia, associated with which, in favourable preparations, is seen a 'stringy' appearance of the cytoplasm. These reflect a tremendous wealth of ribosomes and endoplasmic reticulum. Secondly, there can be seen a prominent, pale-stained area in the cytoplasm near the nucleus; this, the so-called 'juxtanuclear vacuole', contains the Golgi material of the cell. The latter material can be identified by traditional impregnation techniques and of course by electron microscopy (Figs. 34 & 36). A third cytoplasmic feature commonly but not invariably discernible with the light microscope is the presence of small scattered granules or droplets of material giving a positive reaction with the PAS technique.

The nucleus, very rich in RNA, has no particularly specific structural features. There are generally one or two, but often three or more, nucleoli present. The nucleo cytoplasmic ratio is high.

As mentioned on p. 65, osteoblasts associated with the rapid deposition of bone (i.e. whose synthetic activity is high) are especially basophil, plump and polyhedral. In contrast, resting osteoblasts are usually flattened; their cytoplasm is much less basophil, and juxtanuclear vacuoles cannot be seen. Sometimes their cytoplasm is very much reduced in quantity.

2. FUNCTIONAL SIGNIFICANCE

For many years, though everybody agreed that osteoblasts exist, and are found on trabeculae of growing bone, there was sharp disagreement about their role. One view was that they secrete the bone matrix. According to the opposite view, the main upholders of which were Leriche & Policard

(1928), osteoblasts play no part in the secretion of the matrix; on the contrary, they act to confine the forming bone, mould it and keep it in check from spreading too far. However, it is now generally agreed that their role is to synthesize matrix. This opinion is based on evidence of the following kinds.

(a) TISSUE CULTURE

Perhaps the most elegant and convincing of all the experimental evidence that osteoblasts construct bone is provided by some early work of Dame Honor Fell (1932). Small rectangular pieces of bone were excised from tibiae of the embryonic fowl and explanted as tissue cultures, according to the 'hanging drop' technique, in a medium consisting of fowl plasma and embryo extract. The explants soon became surrounded by a halo of cells growing out from their cut edges.

The cultures were then opened and the original bone explants were carefully excised and discarded from the zones of outgrowth, which were thus left with a central gap. The cultured tissue pieces were then washed, trimmed and recultured. The central gap rapidly disappeared, its place being taken by cells dividing and migrating inwards. Subcultures were made at 48–72 hour intervals when the central zones were left undisturbed, but the peripheral outgrowth areas were trimmed back. Some days later, microscopic examination of the living cultures revealed the development of opaque plaques within central areas of the cultures. When fixed and stained, these were seen to be bone, or, in a few cases, cartilage. Thus, cells derived from the bone explants were capable of producing fresh bone on their own. The fact that cartilage sometimes also appeared (Fell, 1933) has an interesting reflection in the healing of mammalian fractures for, in callus, newly-formed bone and cartilage are usually found side-by-side.

In further experiments, the relative osteogenic potency *in vitro* of samples of periosteum and endosteum of embryonic limb bones was tested. Explants prepared from fibrous periosteum, containing fibroblasts but no osteoblasts, though apparently healthy enough in culture, made connective tissue fibres, but neither bone nor cartilage, whereas explants from young, cellular, inner periosteum, rich in osteoblasts, did so. These experiments point very clearly to the osteoblast as the causal agent in bone deposition. This conclusion is of course strengthened by the fact that, in the intact organism, osteogenesis only occurs in the presence of demonstrable osteoblasts.

These pioneer tissue culture experiments by Fell have since been repeated and elaborated many times by others. For one example, the same basic experimental principle has been successfully extended to rodent tissue; the outgrowth zone, instead of being explanted *in vitro*, has been transplanted to the anterior chamber of the eye. There, chondro-osseous tissue formed more quickly and extensively than in tissue culture (see McLean & Urist, 1968).

A second kind of tissue culture was carried out by Fitton Jackson & Smith (1957), using a method previously introduced by Moscona (1952) to dissociate tissues and organs. Pieces of embryonic bone were incubated briefly in dilute trypsin, following which the osteoblasts became loosened, could be detached from the bone by shaking, and then harvested as a matrix-free suspension. Such cells would form 'typical early osteoid' *in vitro*. Similar work has also been carried out with cartilage production by dissociated cartilage cells.

(b) ELECTRON MICROSCOPY

In some respects work with the electron microscope has very much clarified the role of the osteoblast, but it has by no means settled the problem.

Perhaps the chief positive contribution has been to show that the structural organization of the osteoblast is that of a protein-exporting cell. Thus, the cytoplasm of osteoblasts contains large quantities of rough endoplasmic reticulum (ER) forming typical cisternae (Figs. 34, 35 & 36) together with plentiful ribosomes. The cisternae are often much distended in cells considered 'active' from their general appearance and situation (Figs. 37 & 38). It has been reported that electron dense material is located within the cisternae (Cameron, 1963, 1968). Plentiful, free cytoplasmic ribosomes and polyribosomes are also present. It is, of course, the presence of ribosomal RNA which confers the typical light microscopic cytoplasmic basophilia on these cells. A well-developed Golgi apparatus (G) is readily seen in osteoblasts (Figs. 34, 35, 36 & 38) and this corresponds to the pale juxtanuclear area seen with the light microscope. The function of the ER and G is discussed again when the role of the osteoblast is considered in detail (p. 77).

The osteoblast nucleus does not seem to have any unusual features in electron micrographs. The nuclear membrane is double and so-called 'pores' are present, as in many other cell types; the cytoplasm often contains fascicles of microtubules, also observed in cells other than osteoblasts. Mitochondria, naturally, are present. Several research workers have reported on the presence of small electron-opaque 'dense bodies' within the mitochondria; according to one school of thought, their presence affords one link in a chain of evidence to the effect that mitochondria are implicated directly in calcification. This hypothesis is discussed on p. 79.

Although the fine structure of osteoblasts indicates that they manufacture 'export' protein, and this, of course, is collagen, there is no very

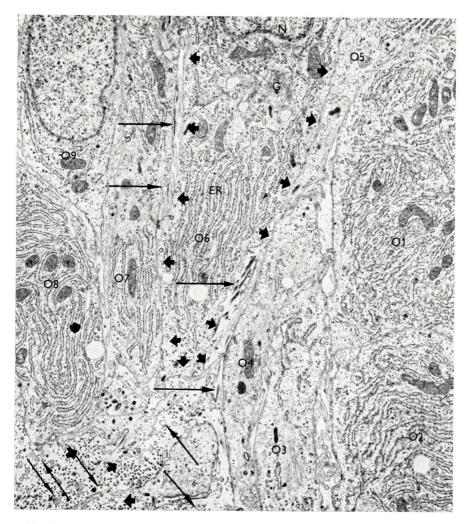

Fig. 34. Embryonic avian bone, fixed in buffered osmium tetroxide. Araldite embedded. Magnification, ×9200. A survey picture. At bottom left corner, the outer edge of a bone trabeculum is present. It consists of newly-formed preosseous matrix; its collagen fibrils are mostly seen in transectional view as rounded profiles. The outermost and most recently formed fibrils (lower oblique black arrows) are widely spaced and of smaller diameter; the older, deeper fibrils are closer packed and larger (horizontal black arrow).

Amidst the newly-formed fibrils can be identified osteoblast processes of varying size, cut in various planes (oblique black arrows). Other processes are visible between osteoblasts some distance from the bone edge. Cross-banded collagen fibrils can be identified near some of them (horizontal black arrows).

The field is mostly occupied by parts of the bodies of a number of osteoblasts (labelled O1–O9). The cell membrane has been delineated by small arrows for O6, and the small portion of its nucleus present in the illustration is labelled 'N'. The Golgi area of this cell (G) is distinctly seen, as are the cisternae of the profuse rough endoplasmic reticulum (ER). Mitochondria are also present. Note the profusion of ER in all the other osteoblasts.

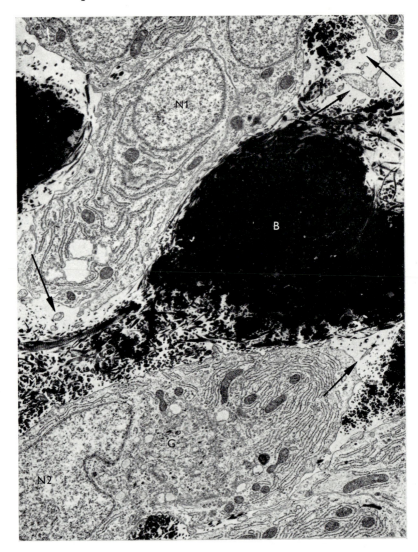

Fig. 35. Technical details as for Fig. 34. Magnification, × 9200. The black mass in the centre (B) represents fully calcified bone; below and to the left, to the margin of the picture, it continues into more recently deposited, incompletely calcified pre-osseous matrix (M) in which collagen is clearly seen. N1 and N2 are osteoblast nuclei. Portions of other osteoblasts are also present. A prominent Golgi area (G) is present near the indented nucleus N2; there is a profusion of ER nearby. Several mitochondria can be seen. Osteoblast processes of varying size are recognizable (oblique arrows).

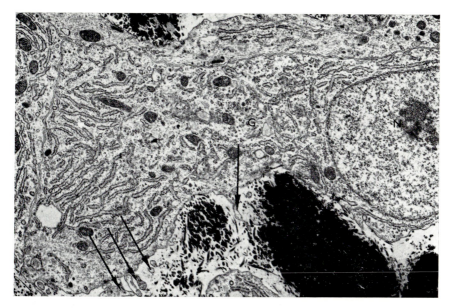

Fig. 36. Technical details as for Figs. 34 & 35. Magnification, ×9200. Fully calcified bone matrix appears black. The nucleus (N) and part of the Golgi area (G) of an osteoblast are surrounded by masses of endoplasmic reticulum. At one point (vertical arrow) a process runs out from the surface; others are seen in oblique and transverse section plane nearby (oblique arrows).

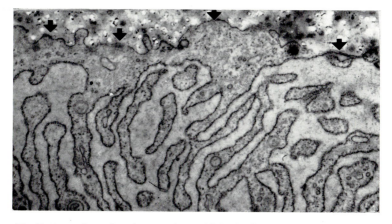

Fig. 37. Embryonic avian bone, glutaraldehyde fixed. Magnification, ×20000. The plasma membrane of part of an osteoblast runs horizontally across the field (arrows). The peripheral cytoplasm beneath contains much-distended ER channels and cisternae which contain electron-dense material. The small, white (empty) circular profiles beyond the free border of the cell are due to artefactual loss of transected collagen fibrils.

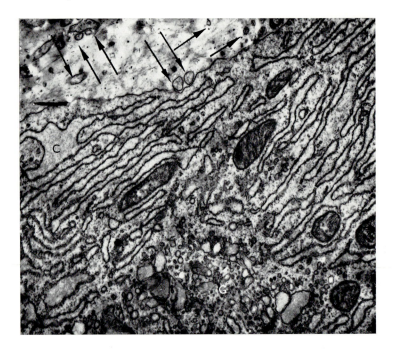

Fig. 38. Embryonic avian bone, fixed in osmium tetroxide. Magnification, ×20000. Part of the cytoplasm of an osteoblast is shown. The rough ER is plentiful and dilated into cisternae (C). Part of the Golgi zone is present (G); its vesicles and smooth lamellae are recognizable. Mitochondria are present. Note also the osteoblast cell processes (oblique arrows).

convincing microscopical evidence that they are involved in the production of the protein–polysaccharide ground substance. However, it is not yet certain whether the latter can even be demonstrated in electron micrographs of bone. It has been suggested (see p. 41) by Balazs (1965) that the role of the ground substance, in certain situations at least, has more to it than a mere space-filling; Smith and colleagues (Smith *et al.* 1967; Smith & Serafini-Fracassini, 1968; Smith & Frame, 1969) claim to have actually identified the ground substance in electron micrographs of rabbit cornea as fine, filamentous material. Electron micrographs of articular cartilage and intervertebral disc material show the presence of a background meshwork of very fine fibrillar material arranged as a meshwork. These fibrillae do not bear the cross-banding characteristic of collagenous fibrils, though of course some of the latter are present, as well as unusual banded structures (Meachim & Cornah, 1970; Cornah *et al.* 1970). Rather similar patterns were seen in electron microscope studies of monolayer preparations of cartilage protein–polysaccharide molecules (Rosenberg *et al.* 1970). Nothing comparable seems thus far to have been demonstrated in bone matrix.

Ground substance is thought to give a positive reaction with the PAS technique (see p. 3). The cytoplasm of osteoblasts contains small PAS-positive droplets (Heller-Steinberg, 1951; Fitton Jackson & Smith, 1957; Knese, 1964). Occasional odd vesicles of about the same size are seen in electron micrographs (Fig. 35). It has been suggested that the two correspond (Dudley & Spiro, 1961). Collagen and 'ground substance' are, of course, found in connective tissue as well as in bone. It is known from work with labelled sulphate (Glucksmann *et al.* 1956; Grossfeld *et al.* 1956), sulphur being an important constituent of 'ground substance', that fibroblasts in connective tissue are involved in the synthesis and secretion of 'ground substance', and of course on other grounds that they are responsible for the collagen. It seems more than likely that osteoblasts behave in the same way.

(c) AUTORADIOGRAPHY

Much valuable information has been obtained from work with radio-labelled compounds. They have been used to shed further light on various aspects of the osteogenic process, such as the way in which osteoblasts handle amino acids, and on cell kinetics. From the latter work it has even been possible to make rough approximations of the amount of bone matrix formed by a single osteoblast per unit of time.

At the light microscope level, it has been known for some ten or more years that, following injection of tritiated glycine, radioactivity is demonstrable over osteoblasts within a short time; soon after this, it disappears from the cells but appears in adjacent newly-formed bone matrix (Carneiro & Leblond, 1959). Young (1962*a*, *b*) found that labelled glycine incorporated by osteoblasts had already begun to be deposited on adjacent bone surfaces within one hour. Labelled proline (the other major amino acid constituent of collagen) behaved (Fig. 41) in the same way; four hours after injection most of the osteoblast content of labelled material had passed to the adjacent bone (Young, 1964). Labelled proline has also been extensively used for studies on collagen formation in cells other than osteoblasts, for example, by fibroblasts in wound healing (Ross & Benditt, 1962, 1965) and chondrocytes (Juva *et al.* 1966). The autoradiographic experiments are interpreted as demonstrating that synthesis of collagen takes place by osteoblasts, fibroblasts and chondrocytes, and by odontoblasts (Carneiro & Leblond, 1959), since other cell types do not concentrate labelled proline or glycine.

Morphological and autoradiographic studies with the electron microscope have carried knowledge further (for example, Ross & Benditt, 1965).

It seemed likely until recently that the pathway of label through the osteoblast followed what was regarded as the classical route in protein-synthesizing cells; thus, it was thought to become located first in the cisternae of the rough endoplasmic reticulum, next in the Golgi area, and finally to appear outside the cell.

Recent quantitative electron microscope autoradiographic work with labelled proline shows that, for cartilage at least, this is not the only route. It was found (Salpeter, 1968) that 'a considerable amount of radio-activity fluxes through the ground cytoplasm and the possibility cannot be excluded that some secretory components leave the cell directly from the ground cytoplasm'. Morphological as well as tracer studies show that the cell's collagenous product is released as a diffusible monometric precursor of fibrous collagen. It seems more than likely that the osteoblast goes about synthesizing collagen for bone in the same way as the fibroblast for connective tissue, but this has never actually been proved to be so. For the fibroblast it is now known that the cytoplasmic polyribosomes form an important site of collagen synthesis. A good review of molecular aspects of collagen synthesis is given by Fernandez-Madrid (1970). It is generally accepted now that the final production unit is the soluble tropocollagen macromolecule, which is secreted extracellularly, where side-to-side and end-to-end linkage of the units takes place to generate visible cross-banded fibrils.

From the cell kinetics point of view, M. Owen (1963) worked with tritiated glycine and thymidine in young rabbits; the former labelled the matrix, whilst the latter labelled cells synthesizing DNA. It was found that the width of the band of newly deposited matrix in bone forming beneath the periosteum of the femur increased by approximately 78 microns per day. It was estimated that 'on an average, an osteoblast produces two or three times its own volume of matrix during its most active period on the periosteal surface'. In subsequent work with labelled uridine (1967) Owen was able to show that the level of cytoplasmic RNA labelling was two to three times greater in 'highly differentiated osteoblasts' than in 'pro-liferating preosteoblasts', which could be correlated with more active protein synthesis in the former than in the latter.

Apart altogether from work on osteogenesis, autoradiography has found many applications in the study of mature bone and these are referred to as appropriate in other sections.

3. SOURCE OF THE BONE CONSTITUENTS

In the preceding sections evidence was outlined to show that osteoblasts can manufacture bone in tissue culture, that their fine structure is that of a protein-exporting cell, and (by means of autoradiography), that they

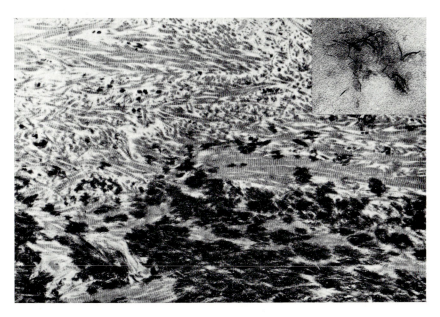

Fig. 39. Embryonic avian bone, glutaraldehyde fixed. Magnification, × 10000. The top half of the field is occupied by the newly-deposited collagen fibrils of the 'pre-osseous matrix'. Their cross-banding can be seen, and they run in various directions; towards the bottom, the fibrils appear larger. In this more mature matrix, mineral deposits are seen as scattered electron-opaque areas. These become more confluent in the deeper, oldest matrix.

The inset at top right shows a small mineral deposit. Magnification, × 200000. The sides of the inset represent 1250 Å by 1000 Å. The dense crystals are 15 to 25 Å wide and appear to overlie a background of less dense somewhat 'plate-like' material of diameter 100 to 200 Å.

concentrate collagen-constituent amino acids. In this section, the morphological evidence about the part played by osteoblasts in osteogenesis will be discussed in more detail, and the origin of other bone constituents considered.

The electron microscope has confirmed and extended the classical observations made by light microscopy. In the very earliest stages of osteogenesis, fine cross-banded collagen fibrils begin to appear near osteoblasts. Such cross-banded fibrils are never seen actually within the cells. This endorses the view that the finished product exported by osteoblasts is the soluble material, 'tropocollagen'. Once outside the cell, tropocollagen macromolecules apparently link up and join end-to-end and side-to-side to form cross-banded collagenous fibres which are insoluble at body pH (see p. 36). The fibrils nearest to the osteoblasts, that is, the youngest, are often finer than the older ones, which suggests that

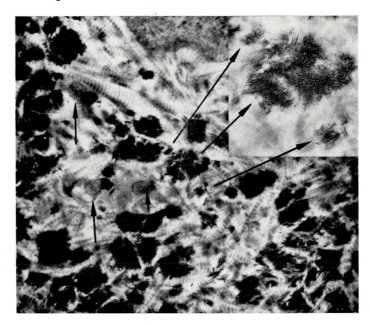

Fig. 40. Embryonic avian bone, osmium tetroxide fixation. Magnification, × 20000. Newly formed, incompletely calcified bone matrix is shown; it consists of a feltwork of collagen fibrils, and includes scattered electron-dense islets of mineral.

The inset at top right shows at higher magnification (× 40000), the islets indicated by oblique arrows. Individual bone salt crystals (which appear as short linear objects) can be seen. Several cytoplasmic processes can be seen in transection view (vertical arrows). These presumably belonged to osteoblasts, and were traversing the matrix.

the earliest-formed fibrils may continue to thicken, perhaps through accretion.

The deposition of collagen fibrils continues, and gradually the thickness of the collagen fibril mat widens to reach dimensions visible by light microscopy. The fibrils tend to run more or less parallel to each other though some interweave and, as the mat matures still further, the fibrils become much more closely packed together (Figs. 34, 35, 39 & 40).

The material produced up to this point corresponds to the 'osteoid' or 'preosseous matrix' of light microscopy, for bone salt crystals are not yet demonstrable in electron micrographs. The width of the osteoid border is variable and is normally very narrow, for calcification commences soon after the osteoid is laid down. Minute aggregates of bone salt crystals appear; they enlarge, and gradually become confluent (Figs. 39 & 40).

Although the electron microscope has provided additional information about osteogenesis and the generation of bone trabeculae by osteoblasts,

innumerable questions remain unsolved. For instance, we remain completely ignorant as to how and where on the surface of the osteoblasts the
tropocollagen macromolecules are liberated; what keeps them sufficiently
near each other, once secreted, for linkages into fibrils to occur and,
equally important, to occur in the appropriate places in respect to the
newly-forming matrix. We do not know how the young fibrils are given
orientation. It is possible that actual movement of the osteoblasts or of
their plentiful fine processes (Figs. 34, 35 & 36) may be involved in pushing
the fibrils around and 'tamping' them into place (Hancox & Boothroyd,
1965). The suggestion that fibres may be spun off from the cell surface was
made by Danielli (1942) long before autoradiography and electron microscopy had been brought to bear on collagenogenesis. The implication of
cell movement as a factor in determining the orientation of fibres has
recently been raised by Boyde & Hobdell (1968, 1969) on the basis of
scanning electron microscope studies of mature bone. Of course, we do not
really even know whether, in fact, osteoblasts *in situ* move actively at all;
it would be exceedingly difficult to arrange a situation in which this
question could be studied in the living condition, and the optical problems
would probably be insuperable. However, the absence of plasma membrane junctions between neighbours makes it seem likely that osteoblasts
do move. There are a number of unanswered questions, too, about the
details of collagen synthesis and fibril formation, which apply equally to
fibril formation by fibroblasts, but these are beyond the scope of this
monograph. Reference has already been made (p. 24) to the way in which,
in lamellar bone, osteoblasts seem to 'remember' to orientate the collagen
fibres parallel to each other.

As mentioned on page 41, the protein–polysaccharide component of
bone matrix has not yet been convincingly demonstrated in electron
micrographs, so that it is impossible to say at what point it makes its first
appearance in newly deposited bone. A similarly non-committal attitude
has to be adopted towards the involvement of osteoblasts in the manufacture of the ground substance and to its delivery and placement amidst
the young fibrils. Cytochemical studies at the light microscope level show
that recently-formed matrix gives a strongly positive PAS reaction very
early in osteogenesis.

The role of osteoblasts (if any) in the calcification mechanism is enigmatical. Mild interest has been aroused in the possibility that their
mitochondria may play some active part. This stems from the fact that
electron-dense bodies are visible within osteoblast mitochondria, as
described originally by Gonzales (1961) and Gonzales & Karnovsky (1961).
Biochemical work subsequently showed that mitochondria can accumulate
large quantities of calcium *in vitro* (Vasington & Murphy, 1962; Rossi &
Lehinger, 1964) and produce large electron-dense bodies within mito-

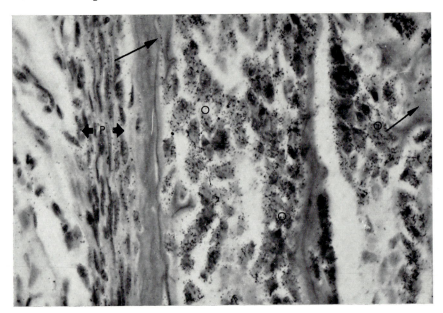

Fig. 41. Autoradiographs of femoral diaphyseal bone of newborn mouse 4 h after injection of labelled glycine. On the left, the fibroblasts of the periosteum (P) are scarcely labelled, but osteoblasts (O) are covered by reduced silver grains. In several places, grains occur over matrix (oblique arrows) suggesting that some incorporation of label has already taken place. Magnification, × 56.

chondria (Greenawalt *et al.* 1964). Cameron *et al.* (1967) administered parathyroid extract to rats and found that their osteoblast mitochondria (at a time when blood calcium level should have been high) possessed dense bodies. Crang *et al.* (1968) reported that the mitochondria of the calciferous gland of the earthworm concentrate calcium. Martin & Matthews (1969) added calcium chloride to fixative for electron microscope studies on calcifying cartilage; they claimed that this made the mito-chondrial dense bodies normally present in hypertrophic chondrocytes even denser. Magnesium had no effect, which suggested a more or less specific calcium-binding effect. They also claimed that [47]calcium could be shown to localize in mitochondria after tissue samples had been exposed to solutions of the isotope *in vitro*, though the evidence presented for this is not very convincing. Shapiro & Greenspan (1969) put forward a theoreti-cal hypothesis that mitochondria are indeed involved directly in biological calcification and Halstead (1969a) considered that such a function might be involved in the production of bone in evolution.

At the present time, the evidence that mitochondria are involved directly in biological calcification is not very persuasive. As Jones (1969) points out, most of the biochemical work seems to have been carried out with liver or kidney cell mitochondria. These organs are chosen for *in vitro* work on mitochondria isolated by cell fractionation techniques because they give a high yield; the preparation of bulk samples of osteoblast mitochondria would be an impossible task. It certainly does not follow that liver or kidney cell mitochondria set a rule which is followed by those of osteoblasts. The weakness of the morphological evidence is that dense bodies are often absent from osteoblast mitochondria (see, for example, the illustrations in this monograph) and, conversely, that they are often present in osteocytes and even osteoclasts (Gonzales & Karnovsky, 1961). Again, the calcareous sponges possess spicules composed of calcite. The spicules are produced by a specific race of cells the calcoblasts, referred to in more detail on p. 96; however, calcoblast mitochondria appear to lack dense bodies so that at any rate, they are evidently not essential for cells to concentrate calcium sufficiently to produce calcite crystals in the metazoa.

8

CALCIFICATION OF BONE

The term calcification, in a biological context, is a wide one. It covers not only the normal mineralizations which take place in the body, as in the preosseous matrix, dental tissues and certain cartilages, but also pathological and ectopic mineral depositions such as kidney stones, the calcification of muscles, blood vessel walls and so on. In this chapter, the term is used in a restricted sense to cover the specialized form of mineral deposition in which salts with a calcium cation and anions primarily of orthophosphate (PO_4^{3-}) and hydroxyl (OH^-) are deposited in the preosseous collagenous matrix.

1. TECHNIQUES FOR THE STUDY OF CALCIFICATION

It is useful at this stage to summarize the techniques available to study the earliest stages of mineralization in bone. The ideal method would allow an initial deposit of solid inorganic mineral to be visualized as soon as possible after it has formed. The size of the deposit at this instant in time is likely to be rather less than 10 Å in any direction (Katz, 1969).

At present one technique alone, electron microscopy, is even theoretically capable of detecting such a particle. In practice, however, it would not be possible to resolve particles of the order of 10 Å size. The specimen must be preserved or 'fixed' in some way on removal from the animal, to prevent changes occurring during the subsequent treatment; it has then to be embedded, usually in a thermosetting plastic, for microtomy, when the embedded specimen is cut into extremely thin (less than 500 Å) sections for electron microscopy. Even with the thinnest possible sections the best that may be expected in practice is a resolution of 20 to 30 Å-sized particles.

With *in vitro* techniques using solutions in which a precipitate is formed, embedding and microtomy etc. are not needed; a drop of liquid may simply be dried down on an electron microscope specimen holder (grid) previously covered by a thin film of carbon. Glimcher & Krane (1968) have reported particles less than 10 Å in size during *in vitro* precipitation of calcium phosphates from solution. No technical details were given, but the basic procedure was probably as described above. A good deal of information has been obtained regarding the crystal species and the size of particles in the earliest stages of precipitation from prepared solutions. It has been possible to deduce that some, if not all, of the initial deposit of mineral in calcifying tissues is relatively amorphous, but in such studies, there has always been some crystalline mineral present, even if only in small quantities; it is not possible to say with certainty which appears first.

Electron microscopy affords a means of directly visualizing the presence of mineral deposits both in sections of tissue and in prepared solutions, but unfortunately limitations of technique mean that there is no guarantee that the deposits are not changed physically, chemically, or spatially by preparative techniques. In fact, it is virtually certain that changes do take place. At the level of resolution under discussion, these changes are likely to be at one and the same time least discernible and yet of the utmost importance. The same limitations apply to a greater or less extent to other techniques (see below) used in the study of calcification.

Only the electron microscope can detect a very small number of particles of mineral deposited in an organic matrix, but, when a deposit exceeds a certain size, electron diffraction becomes an important technique by which the degree of crystallinity of the specimen may be evaluated. In highly crystalline structures it is possible to determine the lattice parameters of the three most symmetrical unit cell types, cubic, tetragonal and orthorhombic. A powder electron diffraction pattern is obtained from the very small specimens under discussion and, by measuring the radii and intensities of the rings obtained, the type of crystal can be ascertained by graphical methods. If preferred orientation of the crystals occurs this can usually be detected and the orientation of the crystals can be related to other structures in a section.

The same specimen can be used for transmission electron microscopy and electron diffraction; thus, it is possible to examine and photograph a selected area of a specimen in the electron microscope and then take an electron diffraction pattern of the same area. Other ways of obtaining electron diffraction patterns from specimens are available, but the selected area diffraction technique permits examination of both the morphology and the diffraction pattern of a specimen. Generally the smallest area from which a diffraction pattern of mineralized tissue can be obtained is about 2 microns diameter. This allows small aggregates of crystals to be examined, but not isolated crystals or the random islets of crystals that occur in the initial stages of mineralization.

Although it is possible to modify high resolution electron microscopes to reduce the size of the electron beam to about 2000 Å diameter, for various reasons related to the specimen and its preparation, the smallest area of almost any biological sample from which a useful electron diffraction pattern can be obtained is of the order of 2 microns diameter.

The X-ray diffraction technique has often been applied to mineralizing tissues. The size of specimen required for this technique is much larger than for electron diffraction, being of the order of 50 microns or more. This effectively rules out the technique for studying the earliest stages of mineralization. Nevertheless, important results have been published from *in vitro* work using X-ray diffraction coupled with other techniques. For example the work of Eanes & Posner (1965) indicates that the first solid which separates out from alkaline solutions of calcium and phosphate mixtures is structurally non-crystalline.

Other experimental methods have been used in the study of calcification usually in conjunction with one or other of the three techniques mentioned above. These have been concerned more with the nature of the chemical bonding between mineral and organic matrix (electron spin resonance) or with the composition and crystal structure of the mineral (infrared spectroscopy). There has been much study of the calcifying ability of physically and chemically modified collagens in which the importance of amino acid side chains and the aggregation of alpha chains and tropocollagen macromolecules has been evaluated.

One other technique, electron probe micro-analysis, has recently been applied to bone research and may become of value in locating the regions where calcium and phosphorus are concentrated in tissues prior to actual deposition of mineral. When used in conjunction with electron microscopy (and generally electron probe micro-analysers are capable of operation as transmission electron microscopes) the micro-analyser is capable of detecting specific elements such as calcium, phosphorus, magnesium, etc. and the amounts of these present in a region as small as 1 micron diameter within an electron microscope section. Thus the mineral concentration in selected zones of an osteocyte cytoplasm, the adjacent osteoid and the nearby bone mineral can be discovered. At present, the size of the electron probe is too large to examine the earliest deposits of mineral selectively, but it is possible that finer probe diameters might become available. Resolutions of less than 1000 Å have already been achieved in specially selected specimens.

2. MECHANISMS OF CALCIFICATION

With the advent of experimental work based on the solubilization and re-precipitation of collagens (Gross *et al.* 1954; Schmitt *et al.* 1955), and of electron microscopy applied to biological material, recent views on calcification have been based primarily on the idea that some of the calcium and phosphate ions occurring in the plasma bathing bone-forming tissue precipitate from solution and form calcium phosphate crystals in and on the collagen fibres. Initially a few calcium phosphate ions arranged in clusters form 'seeds' or nuclei on which crystals of bone mineral grow. The concept of nucleation is widely regarded as being an important contribution to calcification studies, but arguments arise regarding the involvement of biochemical factors in nucleation, and also about the locus of nucleation (Weidmann, 1963).

Collagen forms at least 90 per cent of the organic matrix of bone and it was natural to suppose that it had a major function in calcification. The earliest deposits of mineral in forming bone are very small indeed and visual evidence of nucleation was first provided by the work of Fitton Jackson (1957) and by Robinson and colleagues (Robinson & Watson, 1955; Sheldon & Robinson, 1957) who showed what appeared to be mineral deposits located in specific parts of collagen fibrils. Despite the fact that doubt has been cast on the validity of these findings (Cameron, 1963; Decker, 1966), Glimcher & Krane (1968) have suggested that specific regions within collagen fibrils, 'holes' or compartments about 400 Å × 15 Å, are the sites of initial seeding of the solid phase of bone mineral. This work was based on the Hodge & Petruska (1963) model of tropocollagen aggregation into collagen fibrils. Grant *et al.* (1965) and Cox *et al.* (1967) have recently proposed an alternative explanation of the characteristic band pattern of collagen, based on a less rigid aggregation of tropocollagen macromolecules in the collagen fibril. In accordance with this hypothesis, spaces must occur in the aggregation of tropocollagen macromolecules. These may vary in size from 360 Å × 15 Å to about 2280 Å × 15 Å and be irregularly shaped. These variable spaces would still allow seeding of the mineral phase to occur. Glimcher & Krane (1968) state that approximately 50 per cent of the mineral phase of bone can be accommodated within the spaces in collagen fibrils and argue that this accounts for the radiographic evidence that 60 to 70 per cent of the bone mineral present in fully mineralized bone is rapidly deposited at an early stage in bone formation.

It is possible to examine collagen in the electron microscope after treating it with a negative staining procedure (Horne, 1965) in which 'stain' is thought to penetrate spaces in collagen and also to surround it with a cast of the negative staining reagent. Glimcher & Krane (1968) have shown that collagen so treated looks remarkably similar to calcifying collagen at a very early mineralizing stage, although quite different to the pictures of Fitton Jackson & Smith (1957), Robinson & Watson (1955) and Sheldon & Robinson (1957).

In order to establish that collagen was in fact capable of causing solid mineral to deposit from solution it was necessary to perform experiments *in vitro* with purified protein to ensure that no foreign substance was catalysing the reaction. Collagen may be dissolved in neutral salt solutions and later re-precipitated by dialysis. By repeating this process several times, quite pure collagen can be obtained. Collagens thus prepared from many tissues such as skin, tendon and demineralized bone have been shown to induce crystal formation from prepared solutions of calcium and phosphate ions which otherwise show no signs of depositing crystals (Glimcher, 1959). Various modifications of the normal 'native' collagen structure may be precipitated by altering the conditions chemically and/or physically, but only native type collagen causes mineral to deposit within and on the fibrils.

What causes the collagen to become mineralized? Is it initially a chemical interaction with specific ions or a physical attraction governed by the molecular packing of tropocollagen macromolecules? It has been postulated that certain polar sites on the tropocollagen molecules were arranged in a similar way to the spacings of the atoms in the hydroxyapatite crystal (Glimcher *et al.* 1957; Neuman & Neuman, 1958). This similarity could lead to mineral deposition or nucleation by epitactic overgrowth, that is, the overgrowth by one crystal (the hydroxyapatite) on another (the polar region of collagen) which has a similar lattice spacing. Glimcher & Krane dismiss this as a means of initiating the first deposit of mineral, but recently Marino & Becker (1970) suggested that areas of identical crystal order exist in both bone mineral and collagens from various sources.

Alternatively, chemical binding could occur between the polar side chains of tropocollagen and, for example, phosphate ions in solution, as the initial reaction of nucleation. Solomons & Irving (1958) indicated that there was an interaction between bone mineral and the epsilon amino groups of lysine and hydroxylysine of bone collagen. However, Glimcher & Krane have shown that these results were probably artefactual, due to the barrier to diffusion of reagents formed by the presence of mineral in bone, which prevented interaction. Also, they showed that formation *in vitro* of apatite crystals on collagen fibrils from specific solutions is not dependant on the interaction of any one particular charged amino acid side chain with phosphate or calcium ions in the solutions. They rather believe that the *in vitro* formation of inorganic crystals depends to a large extent on the properties of the solutions of calcium and phosphate used in the experiments and on the overall environment in terms of capillary space and electrostatic charge within the tropocollagen molecules. In any case, it is highly unlikely that the collagen of bone matrix is the only factor which is concerned with calcification. Whilst experiments have apparently indicated that collagen from bone is slightly different to that from non-mineralizing tissues, e.g. less soluble (Glimcher, 1959; Miller & Martin, 1968), Miller & Martin believe that at primary and secondary levels, that is, in the amino acid composition and the alpha chain structures which make up tropocollagen, bone and non-bone collagens are essentially similar. Bone collagen appears, chemically, to cross-link more rapidly, becoming more insoluble than non-bone collagens, but how this is specifically related to calcification is not known.

The view that the protein–polysaccharide complex of bone, the amorphous part of the matrix, is implicated in mineralization has been put forward by many workers. However, it must be noted that most of the results have been obtained from studies of calcification in cartilage, not in bone. Cartilage is known to contain considerable amounts of protein–polysaccharide (Campo, 1970) and the fibrous protein fibrils tend to remain small in diameter, less than 500 Å (Anderson & Parker, 1968), although not necessarily so in adult cartilage (Muir *et al.* 1970), whereas bone collagen can grow to well over 1000 Å diameter (Hancox & Boothroyd, 1964). Because calcification of cartilage is considered by the present author to be somewhat different in mechanism to mineralization in bone, I feel that much of the evidence obtained from these experiments may not be related to calcification in bone.

In fact, bone contains very little protein–polysaccharide. It has been shown that in the mineral-free, dry organic matrix (which comprises about 20 per cent of cortical bone) about 0.25 per cent is chondroitin sulphate A, 0.15 per cent is a sialoprotein, 0.3 to 0.5 per cent is lipid in character, 4.9 per cent is hot-water-resistant protein and 89.2 per cent is collagen. Other, as yet uncharacterized, protein–polysaccharides make up the remaining 5 per cent of organic matrix (Herring, 1964). Herring points out that sialoprotein may be formed and laid down in the matrix during bone formation rather than incorporated from the plasma. As this material is probably a cation-binding agent (Vaughan & Williamson, 1964) it could be effective in forming a pool of calcium ions available for mineralization. It has been shown that part of the protein–polysaccharide component of epiphyseal cartilage rapidly depolymerizes during calcification. Depolymerization would release a pool of calcium ions thereby locally raising the ion product $[Ca^{2+}] \times [PO_4^{3-}]$ and encouraging calcification. Unfortunately there is no evidence of this mechanism occurring in bone. Furthermore, analyses such as the one above appear in general to be taken from relatively mature lamellar bone. Campo & Tourtellotte (1967) give analyses of foetal diaphyseal bone which indicate about double the protein–polysaccharide that occurs in three-month calf bone, but on a whole weight basis this still gives only about 0.73 per cent of glycosaminoglycans.

Similarly, the work of Irving & Wuthier (1968) in emphasizing the presence of lipids in calcifying sites was based largely on the calcification of cartilage in leg bones. In fact the lipid content of cancellous bone becomes no more than 0.2 per cent when their figures are converted to a whole bone dry weight basis.

Currently, it is believed that an interaction between the organic matrix and either the calcium or phosphate ions in the plasma is an important prerequisite to calcification. However, it is an indication of our lack of understanding of the chemical dynamics of bone calcification that some authorities favour a calcium ion to matrix binding as the first step; some favour phosphate ion to matrix binding, whilst others feel that both calcium ion and phosphate ion are equally important at this stage.

Glimcher *et al.* (1964) have shown that reconstituted collagen fibrils will bind inorganic orthophosphate although no stoichiometric relationship was established. Miller & Martin have found only about one mole phosphate per mole of normal bone collagen. That is, bound phosphate is present in bone collagen, but in extremely low amounts. However, only 20 per cent (the soluble part) of total bone collagen, was used in these analyses. It is quite possible that collagen or some other part of the organic matrix may be caused to combine with phosphate by enzymatic phosphorylation. Whitehead & Weidmann (1959) have shown that adenosine triphosphate (ATP) is present in the narrow zone of cartilage cells undergoing calcification, that is, an aerobic mechanism operates in this region, and therefore ATP may be implicated in the calcification of cartilage. Weidmann (1963) is careful to point out that because of 'the many functions of ATP in cellular metabolism its exact role in the calcification process is very difficult to determine'. Despite this, a 'new mechanism' for calcification in skeletal tissues involving ATP has been proposed by Leonard & Scullin (1969), which is cyclical in character. In their theory ATP first complexes calcium and then hydrolyses to pyrophosphate and adenosine. The pyrophosphate further hydrolyses to hydroxyapatite and phosphoric acid and the acid and adenosine combine to recreate ATP.

1. $\text{Ade}-\text{O}-\overset{\overset{\text{O}}{\|}}{\underset{\underset{\text{ONa}}{|}}{\text{P}}}-\text{O}-\overset{\overset{\text{O}}{\|}}{\underset{\underset{\text{ONa}}{|}}{\text{P}}}-\text{O}-\overset{\overset{\text{O}}{\|}}{\underset{\underset{\text{ONa}}{|}}{\text{P}}}-\text{OH}+\text{Ca}^+ \longrightarrow \text{Ade}-\text{O}-\overset{\overset{\text{O}}{\|}}{\underset{\underset{\text{O}}{|}}{\text{P}}}-\text{O}-\overset{\overset{\text{O}}{\|}}{\underset{\underset{\text{O}}{|}}{\text{P}}}-\text{O}-\overset{\overset{\text{O}}{\|}}{\underset{\underset{\text{ONa}}{|}}{\text{P}}}-\text{OH}$

   $\underset{\text{Ca}}{}$

   $+2\text{Na}^+$

2. $\text{Ade}-\text{O}-\overset{\overset{\text{O}}{\|}}{\underset{\underset{\text{O}}{|}}{\text{P}}}-\text{O}-\overset{\overset{\text{O}}{\|}}{\underset{\underset{\text{O}}{|}}{\text{P}}}-\text{O}-\overset{\overset{\text{O}}{\|}}{\underset{\underset{\text{ONa}}{|}}{\text{P}}}-\text{OH}+2\text{H}_2\text{O} \longrightarrow \text{Ade}-\text{OH}+\text{NaH}_2\text{PO}_4+$

   $\underset{\text{Ca}}{}$ $\quad\quad \text{CaH}_2\text{P}_2\text{O}_7$

3. $5\text{CaH}_2\text{P}_2\text{O}_7+6\text{H}_2\text{O} \longrightarrow \text{Ca}_5(\text{PO}_4)_3\text{OH}+7\text{H}_3\text{PO}_4$

4. $\text{Ade}-\text{OH}+3\text{H}_3\text{PO}_4 \longrightarrow \text{Ade triphosphate}+3\text{H}_2\text{O}$

   $\downarrow$ Body buffering

   trisodium ATP

This series of reactions implies that at the region of calcification, the tissue will be acid; a pH value of 6.2 is quoted. In fact, orthophosphoric acid is consumed (8 molecules for every 5 sodium adenosine triphosphate molecules), which, as the acid would be present in some form of phosphate, would tend to raise the pH of the surrounding plasma. An equivalent of 15 sodium ions required for buffering purposes would be freed by the consumption of the phosphate, thus little or no pH change should occur. Leonard & Scullin state that because the ATP is under cellular control, the cells not only control the rate of skeletal tissue mineralization, but the amount as well, by controlling the rate and quantity of ATP produced.

Van de Putte & Urist (1966) are of the opinion that, for calcification, the organic matrix must be exposed to a high concentration of calcium ions or calcium and phosphate ions. The localization of the initial chemical reaction of the tissue with calcium ions determines where calcification will occur. After experiments, Strates & Urist (1969) found that tissue must be exposed both to calcium and phosphate ions to develop much mineral. It is noteworthy that the tendon collagen used by Urist was pre-soaked in relatively strong solutions of calcium ions and/or phosphate ions before mineralization occurred. The relevance of this to normal conditions *in vivo* is uncertain.

Because collagens from a variety of tissues which normally do not calcify, as well as those from dentine and bone, will induce nucleation *in vitro* from prepared solutions containing $[Ca^{2+}] \times [PO_4^{3-}]$ ion products at physiological levels, Fleisch (1964) postulated that some kind of inhibitor exists in the normally non-mineralizing zones which is destroyed where calcification occurs. His experimental evidence indicated that very low concentrations of polyphosphates, or pyrophosphates, of the order of 10^{-6} molar, successfully prevented *in vitro* calcification. Fleisch quotes work by Salo (1950) which indicated that collagens can bind stoichiometric amounts of polyphosphates in the zones which are considered to be nucleating regions ('hole' zones or non-bonding zones of collagen). This would inhibit nucleation of hydroxyapatite in these sites and prevent mineralization. According to Fleisch, for a tissue to mineralize, two essential conditions need be fulfilled: (*a*) the presence of a mineral nucleator such as collagen, elastin, keratin or other fibrous protein and (*b*) the presence of pyrophosphatase to inactivate pyrophosphate.

A drawback to this hypothesis is the fact that all tissues appear to contain pyrophosphatases. Russell & Fleisch (1970) suggest that variations in pyrophosphatase activity may account for the differences between mineralizing and non-mineralizing tissues. This would explain why polyphosphate injections sufficient to inhibit aortic and skin calcification had no inhibitory effect on the mineralization of kidney or bone.

Eanes & Posner (1970) have shown that amorphous calcium phosphate is present in bone to a considerable extent. This material forms only under conditions favouring spontaneous precipitation of calcium phosphate salts. Normal serum levels of calcium and phosphate are lower than this. To account for the amorphous calcium phosphate *in vivo*, a local increase in calcium ions or calcium and phosphate ions would be necessary. This idea has been put forward as a requirement in several mineralization hypotheses (e.g. Robison's (1923) original alkaline phosphate model; van de Putte & Urist, 1966; Campo & Dziewiatkowski, 1963) which refer to calcification of cartilage. Eanes & Posner explicitly involved cellular metabolism in mineralization in the following way. Through an (un-explained) cellularly primed and controlled ion 'pumping' mechanism, calcium phosphate concentrations are raised to the level of supersaturation required for the spontaneous precipitation of amorphous calcium phosphate. A secretion mechanism involving a calcium–phosphoprotein complex and utilizing the Golgi apparatus has been suggested by Taves (1965). This complex on release from the cell would react with extracellular phosphatase, freeing the phosphate to react with calcium and precipitate as an amorphous salt.

The latter next becomes the controlling source of ions for the precipitation of apatite crystals. As it is more soluble than apatite, dissolution of the amorphous calcium phosphate provides the local tissue-fluid levels of anion and cation needed for the extracellular formation of apatite. The collagenous matrix still provides the preferential sites for the primary heterogeneous nucleation of the first apatite crystals. In this model, the primary factor responsible for limiting the formation of apatite crystals to those areas that normally calcify is thus not an enzyme, but a cellularly-derived, amorphous, calcium phosphate. On this view, the cell ultimately governs the entire calcification process.

Pictorial confirmation of the fact that the cell is important in the process of mineral deposition in ossification has recently been published by Bernard & Pease (1969). The value of this work lies in the preparative technique in which the specimens are in an aqueous environment for a very brief period (about two to three minutes) only, as compared with the several hours entailed in orthodox electron microscopic methods. In consequence, aqueous dissolution of minute particles of recently-deposited mineral is minimized. Mineralization is said to begin in loci, which appear to have polysaccharide cores, derived from osteoblasts. These loci are found in the osteoid amongst the newly formed collagen fibrils. The loci are 'buds' or extracellular vesicular extrusions from osteoblasts, and mineral crystals are only found in these vesicles, which may be some distance from the osteoblast. The protein–polysaccharide in the initial calcification loci is thought to provide calcium ion binding sites for concentrating this cation.

After nucleation in the locus, hydroxyapatite grows radially into the organic matrix to form islands or 'nodules' of mineral. Bernard & Pease (1969) state that, next, collagen is 'de-aggregated' in the intermediate zone of the 'bone nodule' and that polysaccharides are reduced in density. Finally many 'bone nodules' grow together by accretion, forming the bone of the fully calcified zone.

Whilst it seems very probable that the osteoblast plays some active role in calcification, the description by Bernard & Pease raises many questions. Their techniques were novel and the results obtained, especially regarding the dissolution of collagen within the so-called 'bone nodules', are contentious. For example, they partially decalcified 'nodules' and then showed that the fibrous arrangement of their centres was relatively disorganized, but state that normally, calcification continues as accretion of nodules to form bone seams. However, it has been shown (Boothroyd, 1964) that thin sections of mineralized bone seams, demineralized on distilled water, have a normal collagenous matrix with no disorganization present; also that specimens of chick bone treated by freeze-substitution and sectioned on to a non-aqueous solvent show a normal collagenous matrix when decalcified on distilled water (Boothroyd, unpublished). The fact that in the work of Bernard & Pease aqueous phosphotungstic acid at pH 1.9 apparently did not decalcify the sections is difficult to understand; it is generally recognized that bone mineral is soluble in dilute acids and even in distilled water in thin sections. It argues either for thick sections in which the embedding medium protected the mineral, or a chemical change in the latter rendering it insoluble in this strong acid.

CONCLUSIONS

From the confusion of publications dealing with calcification in biological systems and particularly those dealing with bone itself, it is now becoming probable that calcification of cartilage and of bone are two distinct metabolic processes with different mechanisms. On the one hand hyaline cartilage, with initially a high content of protein–polysaccharides, must suffer a tremendous depletion of this material before calcification can take place and the cartilage cell is generally depicted first hypertrophying and later degenerating as calcification commences. On the other hand, bone calcification takes place with a very low level of protein–polysaccharides, which alters but little throughout the process, and in the presence of a specialized cell, the osteoblast.

It is probable that collagen plays a large part in the mineralization process, but whether as a passive support for the mineral or as an active nucleator is not clear. It is also likely that the reversible reaction, activation $\rightleftharpoons$ inhibition, controlled to a large extent by the metabolic processes of the cell, is important in the process. This reaction is almost certainly not a function of one inhibitor, but of many 'local factors' which include protein-polysaccharides and enzymes, depending on the nature of the calcifying tissue and its environment. But the one factor which is of prime importance and which governs the whole process in normal bone formation must be the osteoblast.

9

CELLS 'RELATED' TO OSTEOBLASTS

In the animal kingdom there are many examples of races of specialized cells evolved to produce hard, mineralized substances. Some of these cells derive from ectoderm, others from mesoderm, and they produce a range of widely differing materials. Two examples have been chosen for further discussion. These are, first, the mesoderm-derived primate odontoblast, responsible for the production of bone-like dentine and second, the calcoblast, found at the other end of the evolutionary scale in the mesohyl of calcareous sponges producing the skeletal spicules.

1. ODONTOBLAST AND DENTINE

Dentine, as a tissue, somewhat resembles woven bone. It has the same basic constituents, i.e. collagen fibres, ground substances and apatite crystals. It is, however, laid out in a rather different and more regular form. It forms the main constituent of the walls of the teeth. The dentine is bounded externally, either by enamel (above the gum margin), or cementum (below it) whilst internally it is lined by the odontoblasts. Dentine thickens during tooth development because additional matrix is added to its inner surface by odontoblasts. Further dentine ('secondary dentine') may be formed in adult life in response to loss of surface dentine as occurs in caries. For a full and proper account of the histology of dentine, the reader is referred to a textbook of dental histology.

The odontoblasts form a rather regular row of tall, columnar-shaped cells (Fig. 42). Their appearance and shape is much more regimented than that of osteoblasts. It has to be borne in mind that the odontoblasts illustrated in Fig. 42, from a developing tooth, were in a very active state. Odontoblasts in the resting state, as in an adult tooth, have a different appearance; they are flatter and much less conspicuous.

From the outer surface of the odontoblast there extend long thin processes, sometimes branching, which occupy the innumerable long fine parallel dentinal tubules; these are analogous to the bone canaliculi (Figs. 42 & 43). However, recent electron microscopic observations have brought out a major difference; within the dentinal tubule, fine nerve fibres run alongside odontoblast processes (Frank, 1966). This settles a long controversy about the innervation of dentine (Anderson *et al.* 1970).

[93]

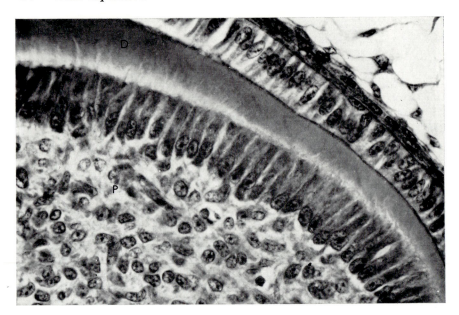

Fig. 42. Decalcified section of part of a developing tooth. Magnification, × 480. The dental pulp (P) is bordered above by a layer of tall columnar-shaped odontoblasts. They have strongly basophil cytoplasm; in some, the pale juxtanuclear area can be seen. Above the odontoblasts is a layer of dentine (D); the most recently formed layer, below, adjoining the odontoblasts, stains paler; it constitutes the 'predentine'. The dentinal tubules can be seen here and there traversing the predentine and on into the dentine. The thin dark line on the outer surface of the dentine (clearly seen towards the right) is enamel, and the enamel-forming ameloblasts lie against it; they are small at this early stage in tooth formation.

In comparison with osteoblasts the odontoblasts are much more regular in shape and their apices are all more or less at the same level.

The contour of the inner edge of the dentine is practically smooth. This indicates that, if odontoblasts make dentine (which they do), the apposition of fresh matrix must take place practically synchronously over considerable distances. In other words, odontoblasts act together at the same approximate rate, moving (or being moved) ahead of the advancing dentine edge, extending their processes as they go.

The role of the odontoblast in dentinogenesis has been studied in more or less the same ways as that of the osteoblast. Morphological studies with the light microscope show a strongly basophil cytoplasm with pale Golgi area; these findings again reflect the presence of profuse rough endoplasmic reticulum with ribosomes, and of Golgi material, as seen with the electron microscope. Tissue culture studies with embryonic tooth germ material, as for bone, demonstrate dentinogenesis by odontoblasts

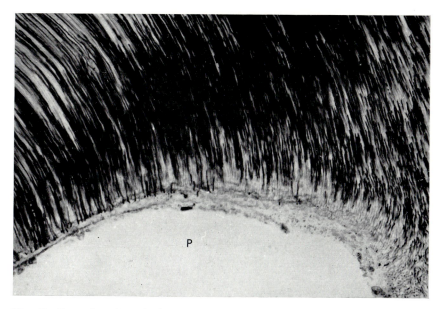

Fig. 43. Ground section of adult human dentine. Magnification, × 120. The curved line running horizontally is the inner border of the dentine. The dental pulp has been lost during specimen preparation; it would have occupied the empty space indicated by P. The myriad dentinal tubules run outwards towards the dentino-enamel junction which was some distance beyond the limit of the field of view.

(Glasstone, 1936). Autoradiography (Carneiro & Leblond, 1959) shows that odontoblasts concentrate collagen-constituent amino acids and transfer them to dentine, reflecting a collagen-producing function similar to that of osteoblasts. It is also thought that they produce the protein–polysaccharide ground substance (Frank, 1966; Noble *et al.* 1962).

Electron microscopy shows that the relatively inactive odontoblasts from teenage teeth possess relatively diminished Golgi material and rough endoplasmic reticulum in comparison with infant teeth (Frank, 1966; Noble *et al.* 1962). This seems analagous to the active and inactive conditions of osteoblasts and osteocytes described in Chapter 6. In adult odontoblasts, however, substantial quantities of glycogen can be demonstrated. The possibility crosses the mind that this may provide energy for maintaining the high areas of membrane surrounding the cell processes.

The electron microscope shows a fundamental difference between odontoblasts and osteoblasts. In the former, the characteristically intimate, side-to-side contact is probably maintained between neighbouring cells because there are desmosome-like structures as well as surface specializations resembling terminal bars near the point of contact of odontoblasts with dentine. They 'form a continuous belt-like attachment between adjacent cells' (Frank, 1966). Perhaps these intercellular bindings restrain individual cell movements and so keep matrix deposition where it should be, i.e. beneath the apical pole. This is probably why the odontoblasts are pushed back together over a wide front, their processes elongating as they move to form a uniform ribbon of dentine. Osteoblasts, in contrast, seem to move about freely and this may be the basic reason for the trabecular texture of newly formed bone.

2. CALCOBLAST AND SPONGE SPICULE

In its simplest form, a sponge consists of a blind-ended tube (or group of tubes) whose wall is penetrated by pores through which sea-water is drawn into the interior, and is expelled at the open end or osculum. The entrancing simplicity of the histological structure of the porifera is fascinating to the vertebrate histologist, though the relatively minute size of sponge cells is disconcerting.

The bodies of the calcareous sponges are stiffened by a 'skeleton' of spicules; the shape, size, and optical characteristics of these vary with species, and with age and anatomical location (Minchin, 1898, 1908; Jones, 1970). Calcareous spicules are largely composed of calcite and have a radiate form, the number of rays being either one (monact), two (diact), three (triact) or four (tetract). Each spicule consists of a single, homogenous crystal of calcite. It is surrounded by a thin, organic sheath.

The outside of the sponge wall is covered by an epithelial-like outer layer, the 'pinocoderm', consisting of a single layer of flattened, polygonal squamous-like cells, the pinocytes. Some are epithelio-muscular, contain myofibrillae, and, on contraction, produce the feeble sluggish movements typical of sponges.

The inside of the tube is lined by cells armed with cilia and long mobile, microvillus-like cytoplasmic tongues. These are the choanocytes or 'collar cells'. The beating of their cilia moves the sea-water into and out of the sponge cavity, whilst their collar of processes snares and enfolds potential food particles. Separating pinocoderm from choanoderm is the thin layer of jelly-like mesohyl which is penetrated by the pores, themselves enclosed each by a porocyte cell (see Borojevic *et al.* 1967, for terminology). The mesohyl, in position, at least, is comparable to mesoderm. It is populated by several distinct races of cells carrying out various functions; they are believed to be derived from large, motile cells, the amoebocytes, which, in their turn, derive from pinocoderm or choanoderm by the apparently simple process of slipping below the surface and moving off to wander freely in the mesohyl.

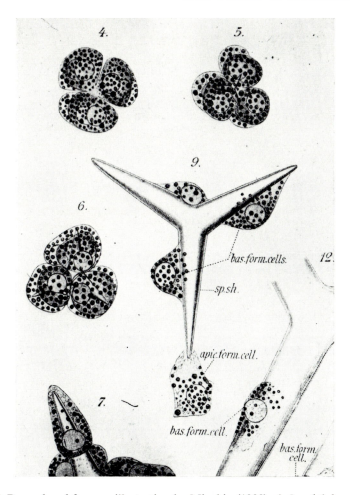

Fig. 44. Reproduced from an illustration by Minchin (1908). *4*, *5* and *6* show stages in the formation of triradiate spicules by sextets of calcoblasts. *9* illustrates appearances at a later stage in the process, whilst *12* indicates the relative size and thickness of a ray in an adult triradiate. *7* shows the relationship of the apical cell cytoplasm to the spicule tip.

Prominent among the dwellers in the mesohyl are the spicule-forming cells, the calcoblasts. The main histological features of spicule formation were described and beautifully illustrated with coloured lithographs by Minchin (1898, 1908). The chief facts, in brief, are as follows.

Each ray of a spicule is produced through the agency of a pair of calcoblasts (Fig. 44). The first inorganic material seems to arise within the cytoplasm of a calcoblast, which soon divides. The daughter cells then

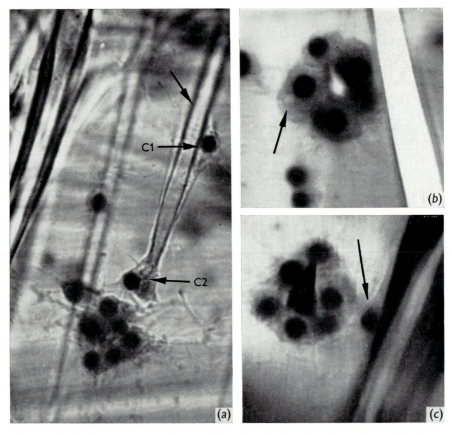

Fig. 45. Whole mount, spread preparation of part of wall of leucosolenia. Osmium tetroxide fixation, stained picrocarmine.

(*a*) A monaxon spicule (oblique arrow) is partly out of focus; the rounded dark nuclei and cytoplasm of its two calcoblasts C1 (the basal cell) and C2 (the apical) are present. The cytoplasm of C1 extends some distance along and appears to surround the spicule shaft. (Compare Fig. 44, *12* & *9*.) It has a darkly stained, round nucleus. C2 envelopes the tip of the spicule; the cytoplasm above and to right of its nucleus contains a clear area. A sextet of calcoblasts lies immediately below and towards the centre of the cell group, in the original preparation, a small quantity of spicule substance could be identified. Magnification, × 2000.

(*b*) A sextet of calcoblasts surrounding an early spicule ray (bright). A pale area can be seen in the cytoplasm of one of the calcoblasts (arrow) (compare Fig. 44, *5*). The spicule to the right of the sextet (and that within) appear bright because partly polarized light was used. These sponge cells are tiny in comparison with osteoblasts and odontoblasts, etc. Magnification, × 2400.

(*c*) A sextet of calcoblasts with one ray of a triradiate spicule (dark, triangular) in early stage of formation (compare Fig. 44, *6*). A monaxon (dark) lies to the right and a calcoblast is present on its surface (arrow). Magnification, × 2400.

migrate to the future base and apex of the spicule (Fig. 44). The former is known as the 'founder', and is thought to establish the shape of the ray. Meanwhile the latter, the 'thickener', adds more calcite to the tip. As the spicule increases in length, the cells separate widely and large areas of spicule surface are left exposed (Figs. 44 & 45). The founder can usually be seen at the base, though the thickener probably moves up and down the spicule shaft. The production of a triradiate spicule calls for three pairs of calcoblasts (Fig. 45).

Until recently it was thought that the calcite was deposited around an organic central axial filament. This would be in keeping with biological calcification systems elsewhere, for as a general rule calcification involves a mesodermal proteinaceous 'skeleton' like collagen for bone or an epithelial protein, such as keratin, for enamel, etc. However, recent work by Jones (1967) has disproved this and it appears that the sponge calcareous spicule is devoid of any kind of axial thread or filament. There is, however, a delicate spicule sheath which may confuse the investigator.

The calcoblast, then, is a connective-tissue type cell whose function is to segregate calcium from sea-water and concentrate it in sufficiently large quantity for calcite deposition to occur. Moreover, it is apparently not involved in the synthesis of an organic product. Such a cell should therefore be an extremely interesting model to compare with the osteoblast, even though the mineral substances with which the two are associated are dissimilar. Calcoblasts, are however, extremely tiny and this limits light microscopy work, so that it is therefore disappointing to have to relate that there is very little electron microscope work upon them. Moreover, sponge cells seem difficult to fix and preserve and the calcoblasts are difficult to section, so that acceptable electron micrographs are scanty at the present time. Though there is rather little reliable evidence as yet, it seems that calcoblasts lack the profuse rough endoplasmic reticulum of osteoblasts, and their mitochondria do not seem to have the electron-dense bodies described for osteoblasts in some cases. Calcoblasts do have a well developed Golgi system; it is just possible that the vacuoles which can with difficulty be resolved in the calcoblast cytoplasm (Fig. 45*b*) may be the light microscope analogue thereof. Future studies on these cells should be most interesting, and tracer techniques would obviously be rewarding. The morphological appearance of the fixed, stained calcoblasts strongly suggests that they must be freely mobile in life, probably moving to and fro along the spicule ray.

10

THE OSTEOCYTE

The osteocyte seems to have become conditioned to life in a rugged environment. In mature, dense lamellar bone these cells are located deeply within heavily calcified surroundings; their cell bodies and nuclei are positioned at relatively enormous distances from blood vessels and must face quite unusual problems in obtaining the necessities of life and disposing of wastes. Even in newly-formed woven bone their difficulties, though less, must still be formidable. It has long been an article of faith among histologists that osteocytes 'communicate' with one another via their processes and, indeed, the microscopical picture shows that osteocytes appear to be 'joined' together via their processes at all levels in osteones. Actual objective evidence for transfer of label along the osteocyte lacunae-canalicular trail, however, is hard to find. One indication is an early electron microscope observation that when bone samples were placed in 'versene' (which decalcifies by chelating calcium) the pattern of mineral loss followed the canaliculi.

The morphology of osteocytes at the light microscope level has already been outlined (p. 63). For electron microscope work, osteocytes and their processes make rather difficult subjects. It is not easy to obtain blemish-free ultrasections of bone and the problem is aggravated because, at soft–hard boundaries such as are formed by osteocyte–lacunae margins, cracks, splits and thickness irregularities are frequently generated. Another disappointment is that the chances of seeing a process neatly sectioned along its axis are remote. Illustrations of canaliculi accompany our article of 1964 (Hancox & Boothroyd). Terminations of one process upon another are rarely encountered.

The osteocytes are of course contained within lacunae; the bone matrix immediately around (in embryonic bone) may be completely calcified or only partly so (Figs. 46, 47, 48, 49 & 50). This obviously reflects, in part, the maturity of the matrix in which the osteocytes lie; in addition, the osteocytes themselves influence the surrounding matrix.

Sometimes the cell membrane of osteocytes appears to fit closely against the lacuna wall (Fig. 49). Sometimes, perhaps more often, there is a gap somewhere between the two (Figs. 46, 48 & 50). Whether the gap is a shrinkage artefact produced by reagents, or whether it is real, is difficult

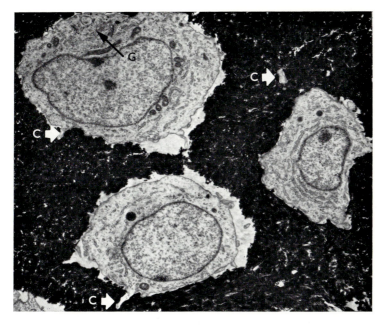

Fig. 46. Low-power electron micrograph (magnification, ×5750) of three osteocytes. Osmium tetroxide fixation. In the uppermost, there is a rather inconspicuous Golgi apparatus (G), and the rough endoplasmic reticulum appears to have been reduced to short channels and vesicular structures. Note the commencement of a canaliculus 'C' containing a process. The bone matrix around the cell appears black where fully calcified. On the right and below there is an apparent gap between cell membrane and lacuna wall. This was actually occupied by uncalcified pre-osseous matrix; into it extends a thin cytoplasmic process which probably connected with the lower cell. The lower left hand cell resembles the upper one in many respects; note start of canaliculus. Above the right hand cell is a short length of canaliculus with process inside (C). The osteocyte to the right has rather more endoplasmic reticulum than its two neighbours.

to decide. It seems quite probable that, like osteoblasts, living osteocytes move a little and change their shape from time to time. Electron microscope appearances can often be interpreted subjectively as snapshots showing short blunt (or finer) processes extending from the osteocyte surface to lacuna wall, stirring up extracellular fluid, or even imparting flow to it along canaliculi. They might then perhaps withdraw and reappear somewhere else on the cell periphery.

The view that it is a fluid (if anything) which separates the plasma membrane from the edge of the lacuna is based on the observation that the gap appears to lack any demonstrable substance (Cooper *et al.* 1966; Cameron, 1963). However, this opinion is not unanimous. According to another school of thought, some kind of ground substance is interposed between

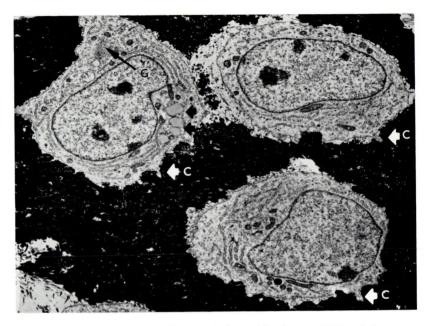

Fig. 47. Low-power electron micrograph (magnification, × 5750) of three more osteocytes. Osmium tetroxide fixation. The upper and lower right hand cells, compared with osteoblasts, contain relatively reduced amounts of endoplasmic reticulum, and appear relatively inactive. The left hand cell possesses rather more endoplasmic reticulum and some dilated cisternae (horizontal arrows); it also has a well-developed Golgi (G) apparatus. This cell therefore presents a more active appearance. The beginnings of several canaliculi (C) can be distinguished.

membrane and wall, and between processes and canalicular wall. This is thought to be a protein–polysaccharide (Baud, 1966, 1968; Belanger, 1969; Dudley & Spiro, 1961).

The appearance of the osteocytes themselves varies between two extremes. At one end of the scale are those which look almost exactly like osteoblasts; these cells have a profusion of rough endoplasmic reticulum, often with dilated cisternae, free ribosomes, and a well-developed Golgi apparatus (Figs. 48 & 50). These structural characteristics suggest an osteoblast-like function. At the other end of the scale are osteocytes having but rudimentary organelles (Figs. 46 & 47). It is tempting to assume that these latter are 'inactive', whilst the former are engaged in producing more matrix. This inference gains some support from the report of Cooper *et al.* (1966) who found that the more deeply placed (and so oldest) osteocytes in adult lamellar bone were relatively poorer in organelles.

The processes of the osteocytes lack organelles; their terminations upon

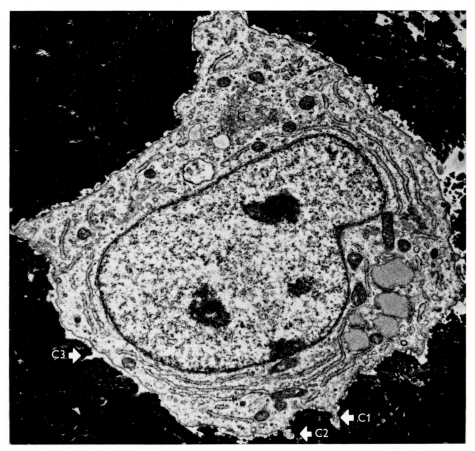

Fig. 48. A Higher magnification (×10150) of the left hand osteocyte in Fig. 47, to show the cytoplasmic organization. Golgi apparatus (G), mitochondria, and rough endoplasmic reticulum channels lie in the cell. 'C1' and 'C2' are canaliculi containing processes; C3 could be a canaliculus from which a process had been withdrawn. There is a small space between cell membrane and bone around the perimeter.

one another are quite simple, and nothing resembling desmosomes or tight junctions has been reported (Cooper *et al.* 1966).

The functions, if any, of the osteocyte remain largely conjectural. This is because they are difficult to investigate by any means other than morphological, and this has inherent limitations, especially the subjective nature of any interpretation of histological pictures. Fine structural attributes, however, provide reasonably objective grounds for supposing that some osteocytes synthesize bone matrix. This supposition is strengthened by their ability to take up labelled glycine (Young, 1962*a*, *b*). As will be seen

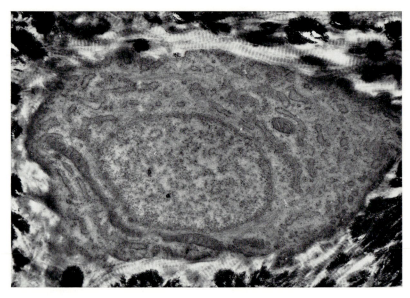

Fig. 49. Magnification, × 15000. An osteocyte is seen surrounded by bone matrix; there is little or no space between the two. The matrix in contact with the cell membrane is mostly immature and incompletely calcified; the cross-banding of its constituent collagen fibrils is clearly visible. The dark areas are islands of bone salt crystals.

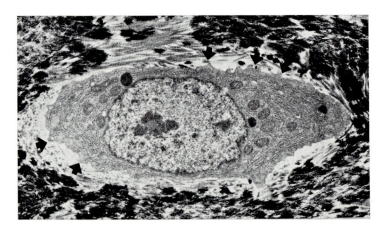

Fig. 50. Glutaraldehyde fixation. Magnification, × 6000. An osteocyte is surrounded by bone matrix which is partially calcified (black areas) and partially unmineralized. The cell membrane is separated from the matrix by a space; the contour of the membrane suggests that it may have possessed processes (arrows), perhaps mobile, in life. The cytoplasm contains profuse endoplasmic reticulum.

in the next section, it is also believed by some people that osteocytes are involved in the converse process of bone dissolution. Whether osteocytes can influence bone is to some extent still uncertain, but, on the other hand, it is known that damage to bone whether by infection, or trauma, leads to demonstrable histological changes in osteocytes. Indeed, one of the criteria used by microscopists in attempting to decide whether a given area of sectioned bone is 'alive' or 'dead' is the gross appearance of the osteocytes. In dead bone, the lacunae look quite empty, the osteocytes having disappeared or shrunk beyond recognition (Fig. 70).

It has been claimed recently (Baud, 1968) that, at the openings of canaliculi into Haversian canals in lamellar bone, special lining cells are demonstrable by electron microscopy. These are said to have two kinds of processes. One sort is thread-like; these penetrate the canaliculi and join up with osteocyte processes. The other sort is more leaf-like; these 'foliar' processes spread between the bone wall and the blood vessels in the canal. The view is put forward that the regulation of metabolic traffic into and out of the 'lacunocanalicular system' is controlled by these cells.

11

LAMELLAR FROM WOVEN BONE

Reference was made earlier (p. 24) to the relative evanescence of woven bone tissue, which, in man and the higher vertebrates, seems to fulfil a temporary role and is substituted sooner or later by the more highly organized lamellar bone. A full account of the substitution process would have to look into such questions as when and where it happens, how, and why. It is beyond the scope of this chapter, however, to go much into times and sites. One example is afforded by the human tibia which at birth is still composed mainly of woven but which, in the ensuing twelve months, is gradually converted to lamellar bone. Another example is provided by post-foetal fracture repair. Fracture callus is a tissue laid down relatively quickly to bridge across broken bone ends and to help immobilize them. It contains large quantities of woven bone. As healing advances, the latter is resorbed, the earlier to go being especially that furthest from the original bone shaft; the woven bone associated directly with the gap persists longer, before it is itself replaced by lamellar bone. Further information on the dates and times is given by Amprino (1963), Enlow (1963), Weidenreich (1930).

Why woven bone is replaced by lamellar remains an enigma. It is, apparently, unconnected with weight-bearing. Ectopic bone induced to form in a non-weight bearing situation, such as beneath the rectus abdominis sheath in dogs, converts from woven to lamellar. In the rather rare human pathological condition, myositis ossificans, woven bone appears in muscles of the arm and is replaced by lamellar; in strongyloides infection in the horse, the same happens in the wall of the aorta. Conversely, limb bones immobilized in plaster following fractures, or those of paralysed patients, do not revert to woven tissue.

The mechanism whereby woven bone is replaced by lamellar involves two processes. Initially, lamellae come to be deposited upon, or near, pre-existing woven bone, often around the periphery of a connective tissue space and oriented with respect to blood vessels (Figs. 51, 52 & 53). Clearly if more and more lamellae were deposited, the space would gradually be filled and so a typical Haversian system would be generated. This process leads to the production of the primary osteones. The second process is the removal of the woven bone substance by osteoclasts. It may be removed

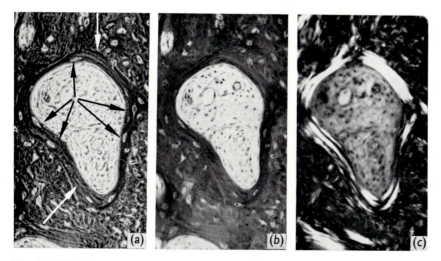

Fig. 51. Celloidin section of decalcified human fracture callus. Magnification, × 180. The same field is shown in (*a*), (*b*) and (*c*). Towards the centre is the oval outline of a vascular channel and this is encircled by a ribbon of lamellar bone; beyond the latter, and extending to the edges of the field, is woven bone.

(*a*) Phase contrast microscopy. Note the flattened osteocyte lacunae arranged concentrically (black arrows) within the ribbon of lamellar bone. Even at this low magnification, lamellae can be distinguished clearly in places (white arrows).

(*b*) Transmitted light. The differences between the woven and lamellar bone are not as obvious as in (*a*).

(*c*) Polarized light. Alternating bright and dark lamellae are clearly seen.

at an early stage before lamellar bone has appeared (Fig. 53*b*) or it may be resorbed, much later, by then enclosed within a covering of lamellar bone. Presumably, the reason for the removal of the last remnants of woven bone by osteoclasts, certainly the result, is the provision of space within the anatomical confines of a particular bone for the laying down of lamellar tissue.

Perhaps the most intriguing enigma of all is why the osteoblasts should suddenly and simultaneously start and continue to generate ordered layers of collagen fibrils instead of interweaving bundles. If cell movement in reality affects fibre orientation then presumably lamellar bone reflects a more regular cell movement track than occurs in woven bone. Unfortunately the nature and mode of action of the factors which, at the microscopic or submicroscopic level, instruct and regulate the osteoblasts, remain to be identified. The substitution of the disordered, haphazard ('straw texture') fibre orientation of woven bone by the more orderly cross-parallel arrangement of lamellar bone has a fascinating and surprisingly close resemblance to events in the wall of the humble

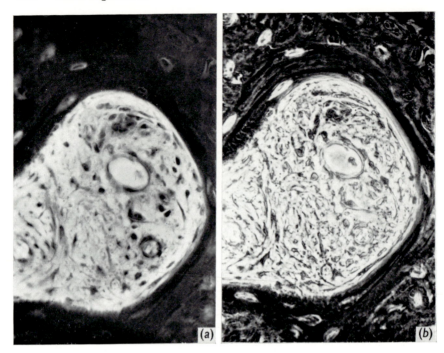

Fig. 52. Part of same field as preceding. Magnification, × 400.
(*a*) Ordinary transmitted light.
(*b*) Phase contrast. The newly deposited lamellae surrounding the vascular space
are quite distinct.

alga *Valonia*. In the young cell, the thread-like cellulose fibres of the wall
lie tangled together, but, in older cells with thicker walls, the fibres are
stacked in organized layers of lamellae whose direction differs by as much
as 90°. Here, the transformation from a simpler, less organized condition
to one of much greater regularity is associated with the maturation of the
cell itself. Another intriguing mystery about the substitution process is
why the osteoblasts as it were 'remember' to construct fresh lamellar
rather than woven bone when secondary osteones are produced following
the resorption of their primary predecessors.

It has long been known that the first Haversian systems laid down
('primary osteones') are themselves removed and the space they formerly
occupied filled up by fresh generations of systems. This constant renewal
is indeed implicit in the term 'bone turnover'. Tomes & de Morgan (1853)
were among the first to grasp the significance of the histological appearance
of bone in this context. In bone sections, appearances very commonly
show how the lamellae in parts of one osteone occupy space which must
have belonged originally to a neighbour, and also 'interstitial' lamellae,

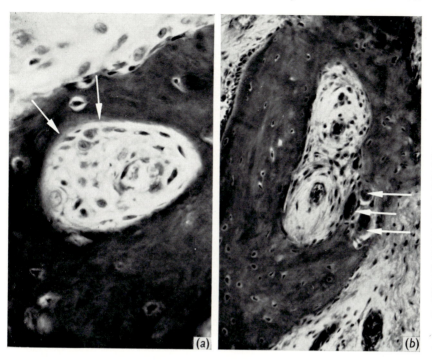

Fig. 53. Celloidin section, human fracture callus. Haematoxylin and eosin.

(*a*) Magnification, ×480. A vascular channel runs through woven bone. Near the top, osteoblasts are beginning to differentiate and a thin film of lamellar bone has been deposited (white arrows). This represents a stage earlier than that shown in Figs. 51 & 52.

(*b*) Magnification, ×180. A stage earlier than (*a*). A channel in the bone has been divided into two by a septum of fibroblasts and collagen fibres; a blood vessel runs through each compartment. The space within each of the latter, now filled by loose connective tissue would probably have become filled by concentric lamellae disposed around the blood vessels; two Haversian systems would thus have formed. A group of multinucleated osteoclasts (horizontal white arrows) is also present. Resorption by them of the woven bone around themselves would very likely have been followed shortly by the appearance of a third vascular channel, itself the potential site for the laying-down of another Haversian system.

which are the last remnants of former complete Haversian systems (Fig. 54). Microradiographs show the same kind of thing; they also add the further information that the oldest lamellae are generally the densest calcified and make it possible to compare, quantitatively, normal and pathological bone (Jowsey *et al.* 1965). The advent of harmless, reliable tracer substances such as aureomycin has made it possible to measure the actual quantity of lamellar bone laid down per unit of time (for example, see Amprino & Marotti, 1964)

Fig. 54. Decalcified lamellar bone section viewed in partly polarized light. Magnification, × 160. Several generations of Haversian systems can be distinguished by the way in which lamellae of some complete systems encroach upon space obviously occupied earlier on by a neighbour (for instance, at H1, H2). 'Interstitial lamellae' correspond to remnants of former systems now lacking central canals (e.g. IL).

In searching for some kind of explanation for the removal of woven bone and its substitution by lamellar, it is tempting to invoke evolutionary concepts. Thus, it seems that ordinary hyaline cartilage proved inadequate as a skeletal building material, and tended to be replaced by calcified cartilage. In its turn, the latter is now in many species substituted by bone, nature having provided for this purpose, in endochondral ossification, one of the most involved and complicated of all histogenetic mechanisms. On this view, the disappearance of woven bone might be regarded as an additional step in arriving at a more perfect building material. Alas, though satisfying teleologically, such an explanation does not really carry us much further in our understanding of the problem.

SECTION 3

BONE RESORPTION

12

OSTEOCLASTIC BONE RESORPTION

Bone absorption or 'resorption', to use the common term derived originally from the German literature, is associated with the activities of the striking-looking giant multinucleated cell, the osteoclast. Resorption by osteoclasts occurs both as a normal event and under pathological conditions.

For example, osteoclastic resorption is part of the normal process of bone modelling during embryonic (Fig. 55) and post-foetal life, and it is responsible for the ordinary slow continuous removal of bone involved in the turnover of the skeleton which proceeds throughout life. In birds, a particularly dramatic example of physiological resorption occurs during the egg-laying cycle (Bloom *et al.* 1941; Taylor & Belanger, 1969). When an egg travels down the Fallopian tube and arrives at the shell gland, there is a sudden need for comparatively enormous quantities of mineral for shell calcification. In the domestic hen, an average eggshell contains about 2 g of calcium, most of which is laid down in the final 16 h of shell calcification; this corresponds to a rate of 125 mg/h. The total calcium circulating in the blood of the hen at any time is 25 mg; hence an amount of calcium equal to the weight of calcium in the circulation is removed from the blood every 12 min (Taylor, 1970). According to Simkiss (1970), 5 g have to be found in 20 h. This enormous demand is met by a very intense osteoclastic resorption of a special form of metabolically-active bone, present in the medullae of the long bones (Figs. 39 & 56).

Abnormal resorption of bone takes place in a wide variety of pathological circumstances. Examples are endocrine disturbances such as hyperparathyroidism either primary or secondary to renal disease; the local effects of tumour cells invading bone (Figs. 57 & 58); and osteomyelitis (Fig. 59). Pressure upon bone leads to its resorption, as when a vascular aneurysm presses on the vertebrae; or in orthodontics when a tooth being moved by means of appliances exerts compressive forces on nearby alveolar bone. Excess vitamin D or insufficient calcium intake may lead to generalized bone resorption. This list could of course be extended several times over, but, irrespective of site and of cause, the basic histological findings are essentially identical. The microscopic hallmark of resorption is the presence of the specific osteoclast cells, together with changes in adjacent bone matrix which can be described, in a word, as

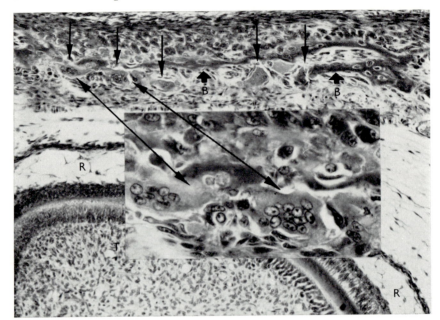

Fig. 55. Decalcified embryonic jaw. Masson trichrome stain. Magnification, × 90. Part of a developing tooth is shown (T). Immediately around it is the reticulum (R) of cells of the enamel again outside which is loose connective tissue. Bone of the jaw is present above in the form of thin trabeculae (for instance, B). Even at this low magnification, large multinucleated osteoclasts can be identified on the side of the trabeculae nearest the expanding tooth (vertical arrows). A portion of the field is shown under higher magnification (× 220) in the centre of the illustration. As well as multinucleated osteoclasts (oblique arrows), typical osteoblasts are visible; see, for instance, at top right corner.

'dissolution'. As will be discussed more fully below (p. 146), in the last decade or so interest has been growing in the idea that osteocytes and perhaps other cells may also be involved in resorption, but there is no longer any serious doubt that the osteoclast is the prime cellular agent.

There have been three main phases in the growth of knowledge about these cells. First came the realization that two entirely different kinds of giant cell ordinarily reside in bone. One, the megakaryocyte, is located in the haemopoietic tissue of the bone marrow. This cell has an enormous, single, lobed polymorphic nucleus. As is now known, it produces the platelets circulating in blood. The other bone-inhabiting giant cell possesses many normal-sized nuclei, and is situated upon or near the bone matrix itself, rather than in the marrow. The first description of these differences was published by Robin in 1849. He was unable, though, to make any suggestions about the possible functions of the two sorts of cell.

The second phase was ushered in when Kölliker of Wurzburg in 1873

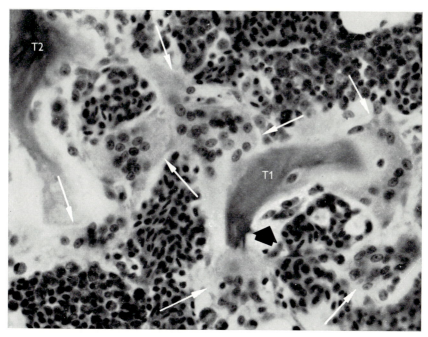

Fig. 56. Fowl medullary bone. Celloidin. Haematoxylin and eosin. Magnification, × 640. Two trabeculae (T1, T2) of medullary bone are present. They are practically surrounded by the cytoplasm of medium-sized, multinucleated osteoclasts (white arrows). A group of three osteoblasts is indicated by a black arrow. The rest of the field is mostly occupied by the closely packed, nucleated erythrocytes and haemopoietic cells of the marrow.

put forward the idea that the multinucleated giant cell actively erodes or 'resorbs' bone and he coined the term 'ostoklast' for it. He founded this hypothesis on the highly suggestive histological picture which points strongly to some effect of cell on bone. He thought that the ostoklast was the universal agent of bone destruction. Kölliker's ideas, based upon the subjective interpretation of morphological appearances, were, however, not by any means universally accepted.

The third phase in our knowledge dates from the period following the Second World War when objective evidence about the function of osteoclasts became available from experiments with such techniques as autoradiography, cytochemistry, electron microscopy, and microcinephotography.

For the sake of convenience, the description of the osteoclast which now follows has been taken up under three main headings; first, general morphology; second, findings with modern techniques; third, its functional significance.

8-2

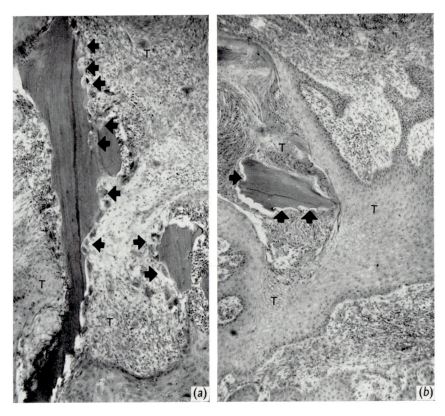

Fig. 57. Human bone. Haematoxylin and eosin. Magnification, ×60. Marrow invaded by secondary carcinoma.

(a) Tumour cells (T) have virtually replaced the marrow, in the greatly expanded vascular spaces. There has already been massive bone resorption. Howship's lacunae containing osteoclasts (horizontal arrows) are present and the 'moth-eaten' contours of the bone edge indicate that active bone erosion was still in progress.

(b) Most of the field is occupied by the epithelial carcinoma cells (T); a small spicule of bone is all that remains. Along its surface can be seen Howship's lacunae (arrows). These now lack osteoclasts, and the inference can be drawn that active resorption had ceased when the specimen was taken.

1. GENERAL MORPHOLOGY

(a) SIZE

Osteoblasts vary greatly in size, but it is practically impossible to measure the bulk of individuals because they usually have an irregular form, with projections and lobes. However, it is apparent that a great many osteoclasts are very large cells indeed, just visible to the naked eye, certainly

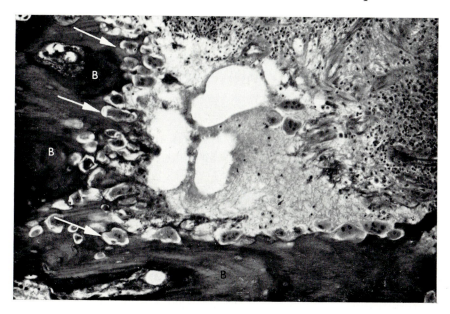

Fig. 58. Human bone. Haematoxylin and eosin. Magnification, × 60. Invasion of marrow by multiple myeloma cells which have replaced the normal myeloid tissue. The edges of the bone (B) are riddled by innumerable erosion lacunae, in most of which can be recognized multinucleated osteoclasts. Some lacunae have been cut obliquely especially at upper and lower left (oblique arrows).

by far the largest cells in the body if one excludes the longest, cylindrical voluntary muscle fibres. The very large osteoclasts have correspondingly large numbers of nuclei, up to three hundred being not uncommon. Medium-sized osteoclasts (such as those arrowed in Fig. 56) contain about 15–25 nuclei per section. They would extend over at least six sections of the thickness shown, and probably contain altogether 50–150 nuclei. At the other end of the scale are much smaller cells, little bigger than macrophages, with two or three nuclei only.

The large masses may give rise to the smaller ones. The evidence for this comes from observations on both fixed and sectioned osteoclasts (Hancox, 1956), and from osteoclasts in tissue culture. In routine sections, one sometimes encounters osteoclasts divided into two or more lobes, each with a quota of nuclei, connected by a narrow waist. With cells living *in vitro*, the large individuals may form two or more lobes connected by a thin filament (Fig. 69). The lobes may wander away from each other and finally separate, or flow together again. Conversely, the fusion of smaller individuals has been observed *in vitro* (Hancox, 1963*b*; Gaillard, 1959). The suggestion has been made, on the basis of the behaviour of

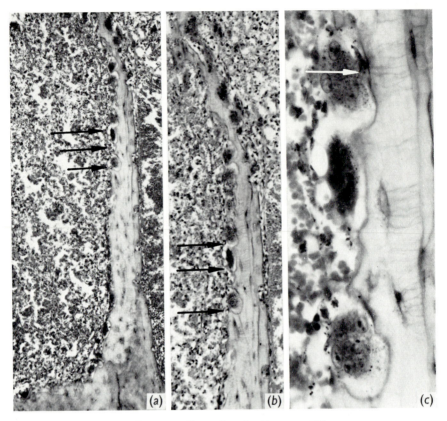

Fig. 59. Human bone. Haematoxylin and eosin. Osteomyelitis.

(*a*) Magnification, ×60. A bone trabeculum extends vertically. At the bottom, its original width is displayed, but a short distance up it narrows abruptly and reduces to a thin splinter at the top. The bone is surrounded by pus and granulation tissue. Its surface is pock-marked by many Howship's lacunae, three of which are arrowed.

(*b*) Magnification, ×100. Portion of above. The same three osteoclasts are arrowed. Others are distinctly seen occupying erosion cavities and bays on the bone surface.

(*c*) Magnification, ×200. The same three osteoclasts. The lower cell occupies a Howship's lacuna; five nuclei can be counted. The middle cell is rather shrunken and dark-staining. This may reflect the onset of degeneration, as mentioned in the text. The upper cell resembles the lower one; note the canaliculi in the subjacent bone. Their parent osteocyte (arrow) now lies outside the bone edge.

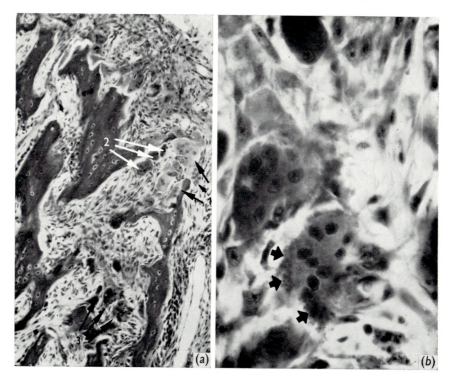

Fig. 60. Embryonic mandible. Masson trichrome.

(*a*) Magnification, × 100. Groups of osteoclasts are labelled 1, 2 & 3. Even at this relatively low magnification, differences are apparent in their cytoplasmic staining density.

(*b*) Magnification, × 400. Osteoclasts of group 3 of Fig. 60*a*. The cytoplasm is intensely stained in comparison with those of group 1 – see Fig. 61*b*. The nuclei are shrunken, irregular in shape, and many are pyknotic in appearance, especially in the lower cell; the peripheral cytoplasm seems, in places, to be breaking up (arrows). It is probable that these are senescent osteoclasts.

osteoclasts *in vitro*, that the organism possesses a pool of osteoclast cytoplasm and nuclei which may shuttle back and forth between one temporary 'cell' and another (Hancox, 1963*a*).

(*b*) GENERAL APPEARANCE

The microscopical appearance of a given osteoclast reflects, among other things, what it was doing at the time it was fixed. Since we know from motion picture studies that they are very highly motile, and may flit from place to place, now resorbing and now resting (Gaillard, 1955, 1959), it is not surprising that they show a great range of histological appearance in

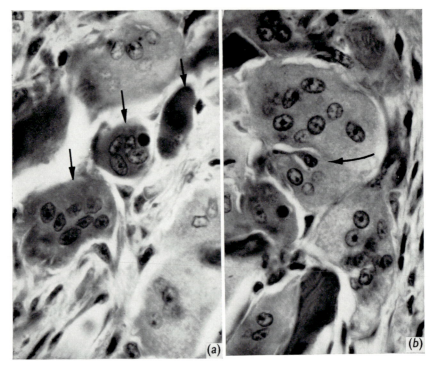

Fig. 61. Magnification, × 400.

(*a*) Osteoclasts of group 2, Fig. 60(*a*). The three cells (arrows) show densely stained cytoplasm. Nuclei of the lower cell are smaller and less regular in shape than those of the middle cell. These osteoclasts can be regarded as, perhaps, showing early degenerative changes.

(*b*) Osteoclasts of group 1, Fig. 60(*a*). Their cytoplasm is relatively less densely stained and is more profuse; the nuclei are more regular in appearance. These cells can perhaps be regarded as normal and healthy in appearance. The horizontal arrow points at a cell (? fibroblast) which though surrounded by osteoclast cyto-plasm has not fused with it.

tissue sections. As shown in Figs. 60 and 61, both cytoplasm and nuclei may vary widely in stain reaction and in general appearance. It is possible that some of the variations may reflect senescence. Similar variations are encountered in the cells comprising the giant-cell tumour of bone.

In routine preparations, the cytoplasm of the osteoclast is often foamy and moderately acidophil; in cells possessing the cytoplasmic chromo-philia and nuclear pyknosis mentioned in the preceding paragraph, the cytoplasm may be intensely acidophilic. With the light microscope, it is possible to see various cytoplasmic vacuoles and mitochondria, but a further cytological description is best deferred to the section below dealing with electron microscope findings.

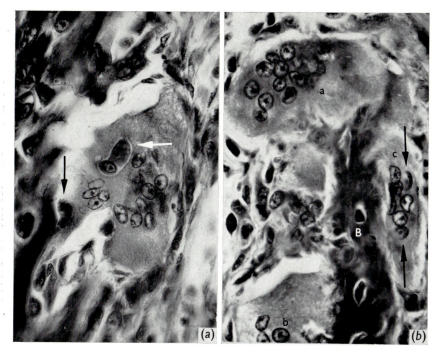

Fig. 62. Embryonic mandible. Masson stain. Magnification, × 200.

(*a*) An osteoclast (badly shrunken by reagents) is present in the centre. About eighteen nuclei are present in this particular section plane. The remains of the ruffled border is indicated by a vertical arrow. The horizontal arrow points to an osteoblast which has become surrounded by the osteoclast.

(*b*) A small, darkly stained bone trabeculum (B) extends upwards in the centre. It contains a few osteocytes. Osteoclasts (*a, b, c,*) invest it closely. Some of the nuclei of osteoclast (*c*) appear concave, when seen 'on edge' (vertical arrows).

The nuclei are characteristically round or oval, most often with a single prominent nucleolus though two are sometimes seen. They look slightly different from osteoblast nuclei, having less chromatin (for example, Figs. 62*a* & 63). When seen in rather thick sections (Fig. 62*b*) they often appear concave. The nuclei tend to be distributed in groups, or more or less all concentrated together, in mid-cell rather than peripherally. This is a convenient point at which to remark that the author has seen neither mitosis nor amitosis in osteoclast nuclei; nor anything which could be interpreted as the fusion of osteoclast with osteocyte or osteoblast.

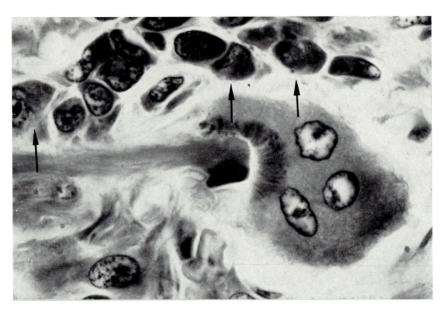

Fig. 63. Embryonic mandible. Masson stain. Magnification, × 1200. An osteoclast (of whose nuclei three are present in this particular plane) is 'wrapped' around the end of a bone trabeculum. The portion of the cytoplasm nearest the bone is differentiated into a 'striated' (or 'brush' or 'ruffled') border. Distinctive osteoblasts can be identified a short distance away (e.g. vertical arrows). Note their darkly staining, basophil cytoplasm; their nuclei contain more nucleolar material than those of the osteoclast. Such a close juxtaposition of bone deposition and bone absorption is a common finding.

(c) ASSOCIATION WITH BONE

On the surface of the osteoclast which is in contact with bone and where resorption is proceeding, a special cytological differentiation is found. This is the 'brush', 'striated', or 'ruffled' border. Osteoclasts are commonly seen in two sorts of 'posture'. In the first, the cytoplasm and brush border embraces or invests the end of a bone trabeculum (Fig. 63). In the second, the cell lies either in a shallow (Fig. 64) or deep (Figs. 58 & 59) depression on the surface of a bone trabeculum. The brush border follows the contour of the subjacent bone surface.

With the light microscope, the border seems to have two components. Nearest to the bone, there are cytoplasmic projections or fringes of varying size, shape, and stain affinity. Some branch and some even look hollow (Fig. 64). Between the proximal ends of the filaments and the main cytoplasmic mass, groups of cytoplasmic vacuoles are present. In some cases it is possible to persuade oneself that the vacuoles lead into channels

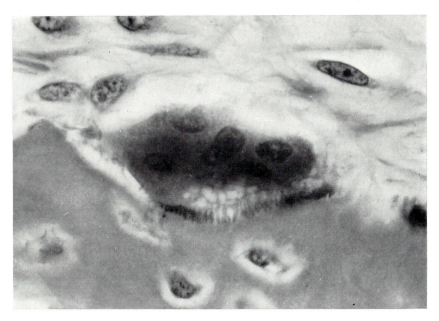

Fig. 64. Embryonic mandible. Masson stain. Magnification, × 480. An osteoclast fits into a shallow Howship's lacuna on bone surface. Its ruffled border can be seen to consist of rather ill-defined cytoplasmic projections or fringes which reach down to the bone. Just proximal to the fringes are a series of empty-looking cytoplasmic vacuoles and it seems that channels between the 'fingers' connect the vacuoles to the bone edge.

which pass between the filaments, and lead to the bone surface (Fig. 64). This, of course, foreshadows the possibility that a to and fro traffic of some kind may go on in this area between the living cell and the adjacent bone edge.

Formerly, controversy existed as to whether any sort of material identifiable as coming from bone could be demonstrated within the striated border or anywhere else inside the osteoclast. The majority opinion was negative (Hancox, 1949). However, the advent of the electron microscope has settled the question conclusively. As will be detailed below, both bone mineral and bone collagen can be demonstrated there.

The morphological evidence can now be summarized as follows. The osteoclast is found in association with the resorption of bone, in health and in disease. It is a multinucleated giant cell of unusual appearance. It fits into pits or wraps around the ends of spicules and where in contact with bone it often has a very striking-looking brush or ruffled border. Whilst these morphological observations with the light microscope strongly suggest (as they did nearly a century ago to Kölliker) that the osteoclast attacks bone, they do not actually prove it. However, within the last decade or so results have been obtained with new techniques which show that osteoclasts are more than passive bystanders.

2. MODERN WORK

(a) AUTORADIOGRAPHY

In a series of classical studies Arnold and Gee at Salt Lake City studied the fate of parenterally-administered plutonium. They showed that this isotope concentrates mainly in liver and bones. They studied its skeletal distribution in the rat by means of autoradiography (1957), and found that it localizes on the outer, newly-deposited surfaces of bone trabeculae soon after administration. The subsequent fate of a plutonium-labelled bone stratum is either to remain upon the bone surface and be covered over (or buried) by new bone if local osteoblastic activity ensues, or, alternatively, the plutonium-binding bone may be resorbed, as part of the ordinary modelling or turn-over processes. They found that when labelled bone was resorbed, alpha tracks in the overlying emulsion of the autoradiographs emanated from within cells which were obviously osteoclasts. Beginning at two hours after injection of the isotope as ^{239}Pu citrate, and increasingly over the next four days, nearly all recognizable osteoclasts became heavily laden with label.

This work was the first in which objective proof was obtained for the passage into the osteoclast of material from adjacent bone surfaces, even though inherent technical limitations made it impossible to say whereabouts in the cell the plutonium was stored.

Much more recently, Owen & Shetlar (1968) have studied the uptake of tritium-labelled glucosamine and found that osteoclasts concentrate it. The significance of this finding is referred to on p. 139.

(b) ELECTRON MICROSCOPY

The much greater resolution of the electron microscope and the infinitely thinner sections needed for use with it have made it possible to obtain very interesting pictures of osteoclasts at work. It is already some fifteen years since the first reports were published by Scott & Pease (1956) and their original observations have since been confirmed and much extended by later reports (Cameron, 1963, 1968; Cameron *et al.* 1964, 1967; Gonzales, 1961; Gonzales & Karnovsky, 1961; Dudley & Spiro, 1961; Hancox & Boothroyd, 1961, 1963; Schenk *et al.* 1967, 1970).

With the electron microscope the chief features of interest to emerge concern the cytoplasmic organization of the cell, and the fine structure of the ruffled border area including the adjoining bone.

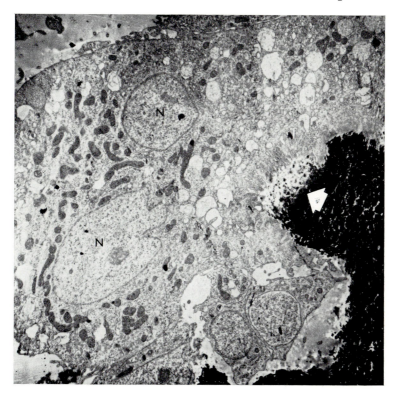

Fig. 65. Electron micrograph. Osmium tetroxide. Embryonic avian bone. Magnification, × 6000. Most of the field is occupied by part of an osteoclast. Two of its nuclei are labelled N. Mitochondria are mainly in the peripheral cytoplasm distant from the bone. The latter is electron opaque due to its mineral and appears black. At one edge however, the outermost bone looks pale (vertical arrow). Here, the osteoclast cytoplasm forms a series of radiating channels and vacuoles leading to the cell interior; this is the ruffled or brush border zone. The cytoplasm between the latter and the nuclei is crowded with vacuoles and vesicles of all sizes and others are present in the periphery.

Low-power electron micrographs (Fig. 65) show that osteoclast nuclei are often indented. The plentiful mitochondria tend to be concentrated in the cytoplasm between the nuclei and the cell border furthest from the bone. In the same area, there are generally a few membrane-bound vacuoles. Sometimes the overlying cell membrane projects as microvilli. Between the nuclei can be identified the stacks of smooth lamellae and vesicles of Golgi material, and a few mitochondria may also be present (Fig. 65).

The cytoplasm in the intranuclear area is characterized by the presence of very large numbers of membrane-bound vacuoles of varying shape, size and, apparently, content (Fig. 66). Description of the ruffled border

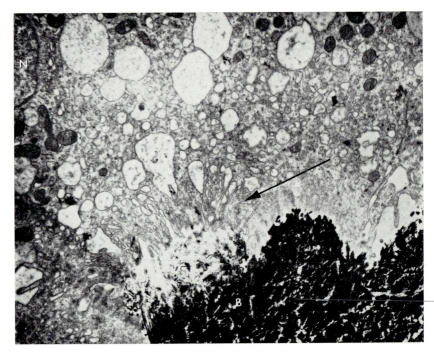

Fig. 66. As preceding; magnification, ×10000. A portion of one nucleus of an osteoclast is labelled N. The ruffled border area (oblique arrow) can be seen as a complex series of cytoplasmic fringes or folds (see Fig. 67), separated by channels which can be seen to lead up into vacuoles, of varying size. The predominant feature of the cytoplasm in this area is the presence of large numbers of vacuoles and vesicles.

area centres chiefly around the folds, fronds or fringes of cytoplasm which extend towards the bone surface, and which are separated by channels running up between them from the bone surface toward the interior of the cell (Figs. 66, 67 & 68). This is an area where cytoplasmic movement goes on very actively and there can be no doubt that many of the channels, vesicles and vacuoles seen in micrographs would, in life, be coalescing, splitting off and shifting about quite rapidly.

Two of the main bone constituents can be recognized within them, bounded by cell membrane and, therefore, extracytoplasmic, if intra-cellular. As shown in Figs. 67 and 68 respectively these are bone salt crystals and bits of cross-banded collagen fibrils.

Changes are demonstrable in the bone near the ruffled border. Some-times, the cross-banding of the collagen fibrils near the surface is clearly visible, as if unmasked by the removal of the mineral deposits which normally encrust the matrix and obscure the banding, or it may be partly

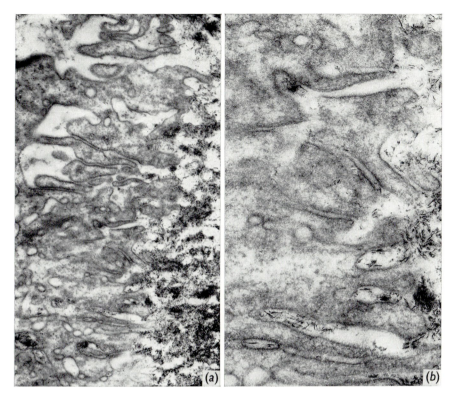

Fig. 67. As preceding.

(*a*) Magnification, × 20000. Part of ruffled border of an osteoclast. Bone edge lies to right, whilst the folds, channels (often branching) and vacuoles of the border are to the left.

(*b*) Higher magnification (× 60000) of part of (*a*). Bone salt crystals are seen as black, electron-dense, needle-shaped rods of varying length, within the folds and channels.

revealed (Fig. 68). Very commonly what look like masses of detached crystals are distinguishable in the border area.

The electron microscopic findings suggest, then, that changes occur in the matrix adjoining the ruffled border of the osteoclast, and that loosened bone constituents are engulfed by the cell. What this means and how it happens will be discussed when certain additional facts, now to be presented, have been described.

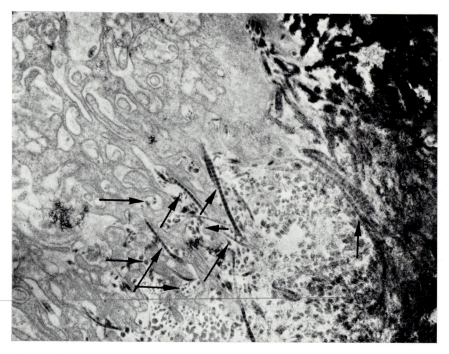

Fig. 68. As preceding. Magnification, × 20000. Bone (B) edge to right, ruffled border in contact with it. The processes, folds and vacuoles of the latter are seen in various section planes. Cross-banded collagen fibrils are visible in longitudinal section in some of the border channels (oblique arrows). In others (horizontal arrows) transected fibrils are present.

At or near the bone edge, in places, cross banding of collagen fibrils is revealed or 'unmasked' by loss of mineral (vertical arrow).

(c) MICROCINEPHOTOGRAPHY

Motion picture studies of living osteoclasts have provided some fascinating data, but for this kind of work it is obviously necessary to use tissue cultures. As a matter of fact, an osteoclast has been observed (but not filmed) once in the living animal; Kirby-Smith (1933) happened to see resorption in a small bone mass growing within an experimental ear chamber he had previously inserted in a rabbit.

Two kinds of observations have been made on osteoclasts *in vitro*. In the first (Gaillard, 1955, 1959; Goldhaber, 1960, 1965), bits of bone with osteoclasts *in situ* were cultured and events recorded microcinephotographically. Bone was seen to melt away beneath the cells, which flit from one place to another, resorbing now here, now there. Though the thickness of the preparations precludes the use of high magnification, the brush

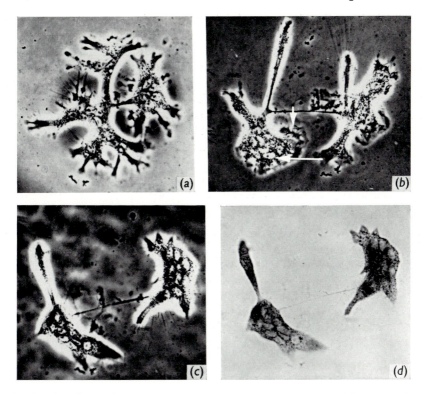

Fig. 69. (*a*), (*b*), and (*c*) are phase-contrast photographs of a living osteoclast isolated in tissue culture. Magnification, × 150.

(*a*) The cell body is drawn out into a number of lobes and processes.

(*b*) Two and a half hours later some of the smaller lobes have withdrawn. Nuclei (horizontal arrow) appear as empty spaces, with the nucleolus as a dark dot. Ruffled border with pinocytotic vesicles can be distinguished in places at the periphery (vertical arrow). Note the very thin cytoplasmic filament, running horizontally which connects the two main lobes.

(*c*) After a further 3 h there have been more shape changes. The cell was next exposed to a 1 in 5000 solution of neutral red in buffered saline for 5 min.

(*d*) Photograph prepared in ordinary transmitted light. The dark cytoplasmic spots correspond to vacuoles staining intensely with neutral red. They are probably lysosomes.

border area can be seen as the seat of a kind of bubbling, 'boiling' activity.

The second kind of observation has been made on osteoclasts which have migrated (by means of a rather characteristic sort of locomotion) out from small pieces of embryonic bone and lie isolated in the medium some little way from the bone itself. Three interesting facts emerge from the study of osteoclasts under these conditions. First, they have a very

energetically active free or undulating border in which vigorous pinocytosis proceeds (Hancox & Boothroyd, 1963). Second, they have great
numbers of cytoplasmic vacuoles which are readily stainable *in vivo* by
dilute neutral red (Hyslop, 1952), which suggests very strongly that they
are lysosomes (Lane, 1968). Neutral red was also used by Barnicot (1947),
who showed that the osteoclasts on the inside of the skulls of new born
mice can be revealed in the living state by vital staining; they can be seen
with the naked eye as minute red spots. Third, as described above, the
living cells produce lobes and extensions of varying shape and size.

(d) HISTOCHEMICAL WORK

The osteoclast is not an easy subject for histochemical studies because
bone sections are difficult to cut without decalcification, a procedure
likely to alter the cells' chemistry; and because bone mineral may compete
for reaction products. These difficulties can be got over by using whole,
unsectioned osteoclasts isolated in tissue culture (Warner, 1964); acid
phosphatase, beta-glucuronidase and leucine aminopeptide were identified
in them. Several other hydrolytic enzymes have been located in osteoclasts
by histochemical methods; a list of them, with references to the workers
responsible for their demonstration, is given in the review by Vaes (1969).

(e) BIOCHEMICAL STUDIES

Obviously the prime difficulty in carrying out work on osteoclasts with the
techniques of biochemistry lies in getting hold of a large enough sample of
material. Though osteoclasts are very large cells, there has not so far been
any test-tube chemical work on them. However, work has been done on
homogenates of skull bones from baby rats (Vaes, 1965; Jaques, 1965;
Vaes & Jaques, 1965) and various enzymes identified and assayed; their
values were compared with those in liver. Whilst one of the short-comings
of histochemical findings is that they cannot at present be expressed
quantitatively, as can biochemical data, it is impossible to say whether the
biochemical findings with bone homogenates refer to the contents of
bone-forming or -resorbing cells, or to fibroblasts, marrow cells, blood
cells, etc.

3. MODE OF OSTEOCLAST ACTION

It is now generally agreed on the basis of the kind of evidence mentioned
in the preceding sections that material derived from bone, i.e. bits of
collagen and groups of (or single) bone salt crystals, are demonstrable
within the folds of the ruffled border and within cytoplasmic vacuoles of

the osteoclast. Traditional stained sections of fixed tissue seen in the light microscope show how the cells occupy lacunae of variable depth (Figs. 58 & 59), whilst motion picture studies (Goldhaber's in particular) reveal how resorption lacunae deepen and alter their contours as resorption proceeds. Now, some ninety-seven years after Kölliker first put the idea forward, the conclusion that osteoclasts play an active role seems to be established beyond doubt. However, several enigmas remain.

First there is the basic question as to how the presumably rock-hard bone edge can be 'dissolved' by the osteoclast. The likeliest hypothesis about this, at the present time, involves both chemical agencies and cellular movement.

The lysosome in recent years has come to be recognized as a new category of cell inclusion which is connected with degradation and breakdown of cells and of intercellular materials (see for example, Dingle & Fell, 1969). The evidence that osteoclasts contain lysosomes is still not quite complete, though very persuasive. For example, there is the electron microscopic finding that the cytoplasm contains innumerable membrane-bound vesicles and vacuoles morphologically resembling lysosomes; according to one report (Scott, 1967) there are 'specific granules', comparable to those in granular leukocytes, which are known to be lysosomal in nature. Then, in the light microscope, there are empty-looking vacuoles in living cells which avidly take up neutral red (Hyslop, 1952) which again suggests they are lysosomes. Live osteoclasts on baby rat skull do the same (Barnicot, 1947). The identification within osteoclasts by Warner (1964) of hydrolases commonly found in lysosomes affords another strong link in the chain. Owen & Shetlar (1968) showed by autoradiography that osteoclasts rapidly take up tritium-labelled glucosamine and use it in the synthesis of a glycoprotein-type material, which can be traced to move, with time, from within the osteoclast to the bone surface beneath. The nature of the substance, whether enzyme or non-enzyme, has not been established, but it seems to be implicated in the resorptive process. There is an interesting correlation between this work and that of Gaillard's with the cine camera. Both suggest that some kind of chemical attack upon, or dissolution of the bone matrix can proceed in an area from which an osteoclast has moved away; as if the lingering effects of some secretory product left behind were still at work. The actual biochemical changes likely to be produced by the action of lysosome-like enzymes on bone tissue are beyond the scope of this account, but have been gone into by Vaes (1968, 1969), together with an account of the effects of stimulators and inhibitors of bone resorption on lysosomal enzymes.

From the point of view of the chemical attack, then, it seems a plausible suggestion that osteoclasts liberate acid hydrolases which degrade the bone. What is more difficult to explain is how these can actually diffuse or penetrate to any depth in the matrix, and whether collagen or ground substance are equally attacked, or in what order, and how the mineral crystals are detached. At all events, if once freed, or if the bone edge became 'frayed', presumably endocytotic and pinocytotic activity by osteoclast would seize upon loosened material (Hancox & Boothroyd, 1963).

A second problem is what directs an osteoclast to a particular site, when to begin to resorb, and when to stop. It is difficult at the present time to see how these kinds of questions, especially in relation to normal bone development, can ever be answered.

A third enigma is why in the course of evolution it has been necessary to develop a multinucleated cell for bone erosion. Why is it better, or more efficient, to have a single enormous osteoclast rather than large numbers of macrophages? It is common knowledge that the latter possess the same kinds of lysosomal enzymes as osteoclasts, and possess an equally active ruffled border. Maybe the answer lies in the possibility that, with its many nuclei, the osteoclast can the more quickly respond to stimuli like parathyroid hormone. It now seems likely (see p. 139) that one of the earliest effects of the hormone is to stimulate the passage of messenger RNA from osteoclast nuclei to cytoplasm, this being followed by a great increase in protein (? enzyme) formation. Less membrane might be needed by a single multinucleated giant cell than a number of small cells.

4. ORIGIN AND FATE

The sudden appearance of large numbers of osteoclasts and their equally abrupt disappearance under both normal and pathological conditions shows that these cells must be formed readily 'on demand'. The first question is whether they arise at, or near, their future field of operation; or are formed at some distant site and arrive at their action station via blood- or lymph-stream. Or perhaps immigrate actively through their own movements. Osteoclasts are very rarely indeed seen anywhere else except in bone, which makes it highly improbable that they are generated elsewhere. The author has once encountered in a tissue section an indisputable osteoclast in the lumen of a blood vessel and there are one or two reports of osteoclasts caught in lung capillaries.

The second question is whether the production of a multinucleated osteoclast involves either mitotic or amitotic nuclear division without cytoplasmic separation, or the fusion of precursors with one or many nuclei. In the absence of reports of nuclear division it is now generally agreed that osteoclasts are formed by the coalescence of precursor or 'progenitor' cells. What remains uncertain is the nature of the latter, which is not surprising in view of the formidable difficulties in identifying them. There are also semantic problems; many of the commonly accepted cytological terms refer to, or are enshrined in, morphological rather than functional attributes.

There are several ways in which the problem of the origin of the osteo-clast can be approached. The oldest, now played right out, is simply to scrutinize stained sections of fixed material in the hope of locating a field where fusion has been caught, as it were, in the act, and in the expectation that the cells fusing might be recognized from their appearance. This has, over the years, proved barren. When at long last a suggestive field is encountered, the interpretation of appearances is a highly subjective matter. One cannot even be certain whether the picture reflects fusion or schism, and morphological criteria on their own are inadequate to define the cells properly.

Another approach is to draw up a list of the special characteristics of the osteoclast, then to attempt to match these with the known attributes of potential precursors and so to arrive at one or two likely candidates. Perhaps the more striking attributes of the osteoclast are its very active ruffled border with pinocytotic tendencies as revealed by electron micro-scopy and microcinephotography, its vigorous motility and tendency to form lobes, and the possession of a well-developed endowment of lyso-somes. Cells of the macrophage series seem to match these requirements better than fibroblasts or osteoblasts. In tissue culture conditions, large osteoclasts seem to tear themselves apart into smaller units which wander away autonomously; smaller units, conversely, tend to coalesce to form large cells. This raises the question as to how one can give a really precise definition of an osteoclast without introducing a temporal reference.

A third alternative, intrinsically the most promising, is to try to label the precursor nuclei or cytoplasm before their incorporation into the multinucleate mass. A great deal of painstaking laborious work has been done using tritiated thymidine as a label. At the risk of gross oversimplifi-cation, it can be said that, soon after injections of labelled thymidine, the first nuclei to label in bone are those within certain relatively undifferen-tiated-looking cells which are not themselves either osteoblasts or osteo-clasts, although they may lie near them. A little later, labelled nuclei are seen within osteoblasts and, within some osteoclasts also, a proportion of the nuclei may be labelled too. It is therefore assumed by some (Owen, 1970) that the cells labelling earliest are the 'progenitors' and that they give rise to osteoblasts and osteoclasts, which thus may share a common ancestor. However, the results of thymidine labelling experiments (which have been repeated many times) can be interpreted entirely differently with equal justification. For one thing, it does not follow that the labelled nuclei seen in osteoclasts were once inside the nearby 'progenitor' cells. They may equally probably have belonged originally to mobile wandering cells or macrophages, which picked up radiothymidine whilst within the blood-stream, and then, after fusion with other macrophages (some labelled and some unlabelled), became a constituent part of an osteoclast. This view is supported by the work of Fischman & Hay (1962). The interesting observation that, following labelled thymidine injections, some of the nuclei within a given osteoclast may be radioactive whilst others are not, has been made on several occasions by different experimenters and has been interpreted in several ways. Young (1962a) used it as evidence for the view that osteoclasts progressively incorporate and lose nuclei, and proposes that osteoclasts may undergo a continual turnover of nucleated components. This adds support to the similar conclusion reached from motion picture studies of osteoclasts isolated *in vitro* (Hancox, 1963b).

Cytoplasmic labelling of potential precursor cells is a difficult task. Jee & Nolan (1963) injected carbon intravascularly and made sections of bone at various intervals. They found that in the early stages, as would be expected, macrophages ingested carbon particles and that osteoclasts already in the bone did not. Some time later, however, indubitable osteoclasts were seen to contain carbon. This suggests very strongly that such individuals must have been formed from fusion of carbon-bearing macrophages. One rather inexplicable aspect of this work, however, is that labelled osteoclasts were not encountered until about fifteen days after the injection. This prolonged 'gestation' period conflicts with most of the evidence which suggests a generation time of hours rather than days.

The story is a complicated one. It would be simpler, perhaps, if the properties of multinucleated cells forming from the coalescence of mononuclear wandering cells did not show one very interesting difference from osteoclasts. The experimental background here is as follows (Kojima & Ogata, 1960). Dried rabbit bone powder was dispersed in carbowax, and samples implanted within the bone marrow through holes drilled through the bones. The animals were injected with trypan blue. Sections showed that the bone powder particles soon became surrounded by multinucleated giant cells of typical 'foreign body giant cell' appearance. Such cells are known to derive from macrophages. They took up trypan blue avidly, as of course do the mononuclear cells. On the other hand genuine osteoclasts appearing in the callus associated with the drill hole did not take in any of the trypan blue. Lack of interest in trypan blue on the part of osteoclasts has been reported several times before. One plausible explanation is as follows. In the ordinary way trypan blue and similar colloids are transported within the cell by endo- or pinocytosis, by macrophages and giant cells; this process seems to go on more or less at any point on the cell surface. Osteoclasts, on the other hand, as is shown by electron microscopy, seem to confine these activities to the brush border area only, which is closely applied against the bone. This being so, failure to take in trypan blue may result not from impotence on the part of the cell but rather because the dye solution is not available to it at the right place.

If at the present time the nature of the osteoclast's precursor remains equivocal, equally shrouded in mystery is its lifespan and ultimate fate. If it is indeed true that the nuclei and cytoplasm of different osteoclasts can exchange, or nuclei enter and leave, then it becomes very difficult to define what one means by 'lifespan' and 'ultimate fate' of these cells. The microscopical appearances do suggest that one or more nuclei of a given cell may undergo degeneration and even on occasion that all nuclei may be so affected (Figs. 60 & 61). The most sensible assumption to make is that an individual osteoclast may get smaller if it loses nuclei (or vice versa), or degeneration may overtake the cell to produce the appearances illustrated.

13

SOME HUMORAL FACTORS
IN RESORPTION

In health, bone is resorbed for two purposes. Firstly, as a necessary prelude to the normal removal and replacement processes through which the tissue of the bones is turned over; secondly, to provide mineral for the maintenance of the blood and tissue calcium at the appropriate level. The extent to which the mineral reserves in the bones need to be drawn upon from moment to moment is a function of many factors. For example, it depends on the adequacy of the dietetic intake and the efficiency of intestinal calcium absorption, itself affected by several variables including vitamin intake; on the rate of utilization and excretion of calcium, or its non-availability for other reasons. A full account of the mechanisms of calcium homeostasis is beyond the scope of this monograph. Many good reviews have been published; those by McLean & Urist (1968) and Vaughan (1970) are recommended. Presiding over these events, of course, are the hormones of the parathyroid glands and the 'C' cells of the thyroid whose mode of action will shortly be discussed more fully.

In abnormal conditions, bone resorption may be the result of factors acting locally, such as pressure upon a bone, infection or invasion by neoplastic cells. Generalized resorption occurs as a result of chronic dietetic mineral and vitamin insufficiency, or to over-activity of the parathyroid glands which itself may be due to neoplasm of the gland, or secondary to renal disease. It may be caused by other endocrine disturbances, and there are a host of other causes, some well understood, others ill defined. In the latter category may be mentioned neuropathic conditions such as acrodystrophic neuropathy (Spillane & Wells, 1969) and Sudek's bone atrophy (Collins, 1966).

This is a convenient point to remark that the role, if any, of the nervous system in the deposition and resorption of bone is not understood; and, indeed, the actual distribution of nerve fibres and endings, both effector and sensory, is rather obscure. The periosteum is said to be sensitive, and contains nerve fibres (twigs of afferent somatic nerves) and nerve endings of Paccinian corpuscle type. Within the bones, nerve fibres are almost confined to the walls of blood vessels though there are some reports of them running free in Haversian canals. Small unmyelinated nerve filaments have been seen with the electron microscope lying outside Haversian blood vessels; enclosed within Schwann cells, they contain dense neuro-filaments. No nerve endings were identified (Cooper *et al.* 1966).

The presence of a nerve supply is not essential for either resorption or deposition of bone, as is clear from the fact that both can proceed perfectly well in tissue culture. Further pointers are that experiments (admittedly on frogs not mammals, but this seems without significance) have shown a practically normal development of the skeleton following complete removal of spinal and sympathetic innervation, and the growth of limb rudiments to virtually normal size and shape when grafted to the chorioallantoic membrane (Murray, 1936). Such rudiments are, of course, completely severed from the nervous system and, as the chorioallantois lacks innervation, must remain outside nervous control. Later work by Bradley (1970) has shown, however, that whilst the bone tissue appears perfectly normal, skeletal muscle of the transplanted limb rudiments is abnormal.

From the experimental point of view, generalized resorption can be provoked in various ways. Firstly, by procedures which result in the depression of the blood calcium level and thus stimulate the parathyroid gland. Peritoneal lavage is an example; if the peritoneal cavity is repeatedly washed out with calcium-free fluid, calcium leaves the blood to the peritoneum. In an evident attempt to restore the level, bone resorption mediated by the parathyroid glands ensues (Talmage, 1967). Nephrectomy also stimulates endogenous parathyroid secretion. The simplest, most predictable and best known procedure is, however, to inject parathyroid hormone either in its purified form or more cheaply as parathyroid extract (PTE). The events following such treatment will next be considered.

1. MODE OF ACTION OF PARATHYROID HORMONE

It has been known for many years that following an injection of parathyroid hormone two sets of changes can be demonstrated. The blood calcium level rises sharply with an accompanying rise in renal phosphorus excretion; and there are cytological changes in the bones which include histological changes in the matrix, alteration in the appearance of osteoblasts and in the number of osteoclasts. It became known more recently that the hormone also acts to influence the calcium uptake of the gut.

In the earlier periods of the history of parathyroid physiology the dual effect on bone and kidney led to opposing views as to which of the two is the primary target for the hormone. It was widely believed at one time that its basic effect was to lower the renal threshold for phosphorus excretion; in order that the Ca/P ratio should remain constant in the face of a marked phosphorus leak, a secondary rise in blood calcium ensued. At that time it was also considered that the rise in blood calcium took place before any cytological changes occurred.

However, it was suspected by many that the hormone had, at least in part, a direct effect upon the bone tissue and its cells. This was proved by ingenious experiments in which whole parathyroid glands were stuck to skull bones with plasma clot and then transplanted (Barnicot, 1948; Chang, 1951; J. J. T. Owen, 1963). An intense resorption of the bone took place near the gland; control tissues had no effect on the bone. Gaillard obtained similar results with tissue culture methods (1955, 1961).

In the experience of many research workers, the rise in the blood calcium level following injections of the hormone, or of parathyroid extract (PTE), began and reached a high (though not the highest) level before any obvious increase could be demonstrated in the osteoclast population. On the other hand, early and pronounced changes are demonstrable in osteoblasts. They lose their plumpness and basophilia to become fusiform in shape (Heller *et al.* 1950; Heller-Steinberg, 1951; McLean & Bloom, 1941; Young, 1963). Their collagen-synthesizing activity declines, as estimated by autoradiographic studies of their uptake of tritiated glycine (Young, 1964). Electron microscopy revealed three kinds of changes in osteoblasts. First, a swelling and disruption of their mitochondria; second, disappearance or disruption of the rough endoplasmic reticulum; third, the appearance of lysosome-like bodies (Cameron *et al.* 1967). There seemed to be no effect on osteoclasts.

In the absence of demonstrable effect in osteoclasts, and since it seems unlikely that the osteoblasts could erode bone, changes have been sought for in other bone cells. It was reported that, under the influence of PTE, the osteocytes undergo alterations; that they accumulate lysosomes (Baud, 1968; Belanger *et al.* 1966) and that the lacunar wall shows evidence of osteolysis (Baud, 1968; Belanger & Migikovsky, 1963). However, it is just as likely that the changes seen in osteocytes are toxic and represent early signs of the death of the cells. This also goes for the changes seen in osteoblasts.

Not all reports ruled out an early increase in the osteoclast population (or changes in the appearances of the cells) which might signify a direct effect of PTE upon them, and so explain the rise in blood calcium level. For example, Young (1964) found that the proportion of 'active' to 'inactive' cells increased, and they grew appreciably larger.

The effect of PTE upon osteoclasts has been much clarified through recent, extremely interesting work carried out by autoradiography. The starting point may be said to be the discovery that the action of PTE in initiating bone resorption depends upon protein synthesis. This was extended in the last few years to observations that PTE alters RNA synthesis in bone cell homogenates. The work of Owen and her collaborators (Owen & Bingham, 1968; Bingham & Owen, 1968; Bingham, Brazell & Owen, 1969) with autoradiography has made it possible to study the effect, not merely upon cell homogenates, but upon the actual different bone cells. The main findings are as follows.

First, following an injection of PTE, the uptake of labelled uridine was depressed in osteoblasts, but stimulated in osteoclasts; increased uridine labelling was first seen ($1\frac{1}{2}$ h after injection) in the nuclei, then 12 h later in the cytoplasm, where it reached a level 200 per cent greater than in controls. This indicates a great increase in nuclear RNA synthesis and transport to the cytoplasm under the influence of PTE.

Second, the increase in cytoplasmic RNA in the osteoclasts was shown to be accompanied by a rise in the utilization of labelled leucine and glucosamine; the latter is used by the cell in the synthesis of a glycoprotein-type material apparently associated with resorption. These findings indicate a stimulation of cell-product manufacture. Among the latter may well also be enzymes, perhaps for primary lysosomes; glucuronidase for instance is a glycoprotein.

Third, the effects on osteoclasts can be correlated, in time, approximately with the rise in the plasma calcium level, which is therefore presumably due to an effect upon those osteoclasts already present in the bone. The osteoclast population does not begin to increase until some hours later, which suggests a secondary effect.

Fourth, the effect of PTE on osteoblasts is to depress RNA synthesis. This confirms earlier work with glycine, and correlates with light and electron microscope observations as loss of cytoplasmic basophilia and rough endoplasmic reticulum.

This very elegant work has carried our knowledge of how and where PTE acts in bone very much further forward. It provides, too, an excellent example of how, with modern methods, problems can be attacked which not so many years ago would have seemed, like putting a man on the moon, utterly beyond our reach. The picture is still not yet complete. The significance of the histochemical evidence of depolymerization of matrix around osteocytes after PTE injection, described by Heller *et al.* (1950; Heller-Steinberg, 1951) remains enigmatical; the precise intracellular target of parathyroid hormone remains to be identified.

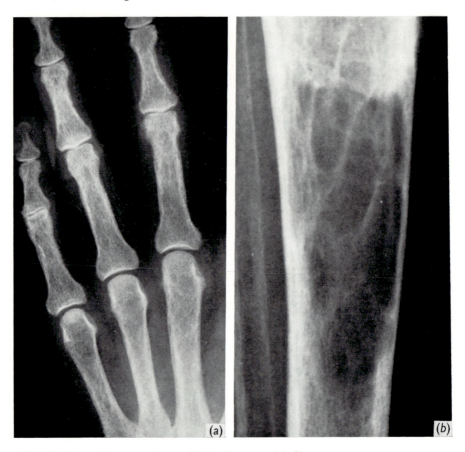

Fig. 70. X-rays from a patient suffering from osteitis fibrosa.

(*a*) Fingers. The phalanges show uneven contour and rarefaction; the terminal phalanges present a 'moth-eaten' appearance.

(*b*) Tibia, mid shaft; there are large cystic areas of rarefaction, and these involve the cortex as well as the medulla.

2. HYPERPARATHYROIDISM IN MAN

The identification of the parathyroid as a gland separate from the thyroid was first made by Sandström of Uppsala (see English translation by Seipel & Hammar, 1938). In 1877, whilst still a medical student, he found a new gland when dissecting a dog and by superficial microscopical examination showed its structure to be quite different from that of the thyroid. During the winter of 1879–80 he carried out further research. Although he identified the parathyroid in horse, ox, cat and rabbit, his main interest was in its occurrence and microscopic structure in man. He

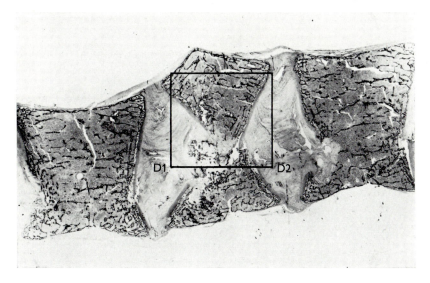

Fig. 71. Figs. 71–74 illustrate changes in the vertebrae of a patient who had suffered from intense hyperparathyroidism secondary to renal disease. Fig. 71 shows a section which includes the bodies of three vertebrae and two intervertebral discs (D1 & D2). Magnification, × 1.2.

The bone trabeculae of the vertebrae are greatly diminished in number and thickness; they appear as thin black lines. The vertebral body in the centre (between D1 & D2) was so weakened that it collapsed. This is reflected in its obvious compression and in the disorganization of the bone trabeculae between the discs, especially in the lower half of the illustration.

published his anatomical and histological findings in 1880. Research into the possible functions of the gland very soon started, but of course the experimental work was carried out on animals, and most modern knowledge comes from animal work and tissue culture studies.

Nature, however, sets up parathyroid experiments in man, particularly in the form of hypersecretion. This may be primary, in the sense that the secretion of hormone is abnormally high because of overgrowth of the gland tissue itself, as happens with adenoma or, more rarely, carcinoma; or it may be secondary, as when parathyroid secretion is stimulated in response to distant changes caused by renal disease.

(*a*) PRIMARY HYPERPARATHYROIDISM

The symptoms of and post-mortem changes produced by a generalized disease of the bones was reported in 1891 by von Recklinghausen as osteitis fibrosa cystica, though, at that time, the fact that it was due to enlargement of the parathyroids was not known. The first deliberate

Fig. 72. Higher magnification (× 4) of the part of Fig. 71 included within square. Below (lowest horizontal arrow) the intervertebral disc (D1) has been invaded by bone trabeculae. Above (upper horizontal arrows) the trabeculae are disorientated and microfractures are present.

removal of an enlarged human parathyroid was carried out by Mandl (1925) of Vienna but it was not until 1931 that the connection between von Recklinghausen's disease and overactivity of the parathyroid was firmly established by Hunter & Turnbull.

Patients suffering from primary hyperparathyroidism show many of the same basic abnormalities as are seen in experimental animals injected with PTE or purified hormone; *inter alia*, there are bone changes and a high blood calcium level. Whereas much of the animal work is concerned with the acute effects of large doses, in von Recklinghausen's disease the condition is generally a chronic one, so that fairly advanced, widespread changes may take place before a diagnosis is made. In cases where the blood calcium level has been much elevated, secondary pathological changes may be seen in the kidneys, where calcification and pelvic stones commonly occur; calcification may also take place in the walls of the arteries, the lung alveoli, and elsewhere.

The changes in the bones in some ways resemble those seen in experi-

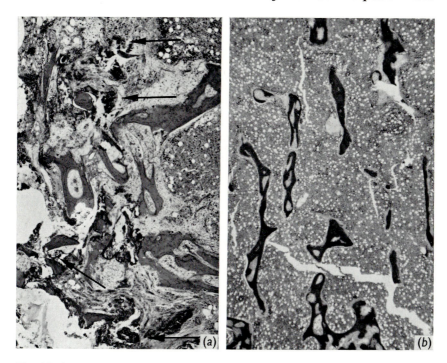

Fig. 73. (*a*) Higher magnification of part of Fig. 72 adjoining uppermost arrow. Broken and displaced fragments of trabeculae are indicated by oblique arrows. Amorphous bone remnants (horizontal arrows) indicate sites of previous micro-fractures. Magnification, × 100.

(*b*) Bone marrow and trabeculae from another region. The trabeculae are much altered; they comprise calcified bone (darker staining) and uncalcified connective tissue (paler staining). (See also Fig. 74.) Magnification, × 100.

mental material. There is a massive osteoclastic resorption; this occurs mainly in the cancellous bone, but the process spreads from medulla to cortex and, on X-ray, there are cyst-like lesions which seem to erode the cortex itself (Fig. 71). X-rays also reveal cyst-like areas of local rarefaction, in bones such as clavicle, ribs and mandible. Histologically these cysts are in fact seen to comprise tumour-like aggregations of fibroblasts and osteo-clasts in a vascular framework. Removal of the affected parathyroid restores the bones practically to normal.

Another characteristic and interesting finding is the presence of large numbers of active osteoblasts forming massive amounts of newly-formed woven-type bone, rather patchily-calcified, and itself often under attack by osteoclasts. Some recent experiments by Marks (1969) bring to mind the possibility that the newly-discovered hormone calcitonin may be involved in this phenomenon. He found that in newborn mice, a small

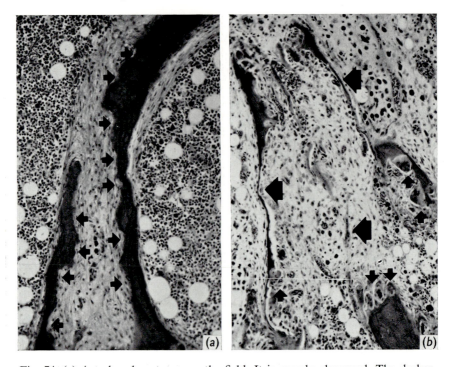

Fig. 74. (*a*) A trabeculum traverses the field. It is grossly abnormal. The darker-staining, calcified bone shows numerous erosion lacunae (horizontal arrows). Osteoclasts are no longer present in the lacunae which indicates that resorption had 'burned out' by the time the specimen was taken. The place of the eroded bone has been taken by uncalcified cellular connective tissue (pale stained). Magnification, × 160.

(*b*) Another area showing the end result of an extreme degree of resorption. Osteoclasts are present here and there (vertical arrows). The trabeculae have been eroded down to the merest wisps in places (horizontal arrows). Magnification, × 160.

dose of PTE given daily for long periods produced a condition of bone trabecular overgrowth ('osteopetrosis'), and that hyperplasia (up to eighteen-fold) of the calcitonin-producing thyroid parafollicular cells ('C' cells) occurred. The suggestion was made that the elevated blood calcium stimulated the C cells to produce more calcitonin which in turn caused osteoblasts to lay down fresh bone.

(*b*) SECONDARY (RENAL) HYPERPARATHYROIDISM

The regulation of the blood calcium level involves, of course, both the kidneys and the bones and there are a number of pathological conditions of the former which produce secondary hyperparathyroidism with advanced bone changes. These changes may be very severe indeed. For example, Figs. 71–74 illustrate changes in the vertebral bodies from a young woman who died of renal disease, and in whom the parathyroids were found at post-mortem to be greatly enlarged. A very severe resorption of bone had occurred. The bone trabeculae were extremely thin and greatly reduced in number. Several vertebrae were so weakened that they had suffered compression fractures, i.e. collapsed.

For a full description of the bone changes seen in pathological conditions, textbooks such as those by Collins (1966) and by Aegeter & Kirkpatrick (1963) should be consulted.

14

RESORPTION WITHOUT OSTEOCLASTS

Though it is now generally accepted that osteoclasts erode bone, it is also thought by some authorities that other kinds of cells may do the same thing. In the past, the macrophage and the vascular endothelial cell have been thought of as perhaps capable of bone resorption. At the present time the osteocyte and the mast cell seem to be the strongest candidates.

1. OSTEOCYTES AND BONE RESORPTION

The idea that osteocytes may cause a dissolution of the matrix surrounding them was apparently first put forward in 1881 (by Rigal & Vignal, according to Belanger, 1969). It was revived by von Recklinghausen in 1910. He coined the term 'thrypsis' for the melting away of matrix around swollen ('oncosis') osteocytes.

In the last decade or so, interest has been aroused again in the possible intervention of the osteocyte in resorption. The work of Belanger and Baud and their associates is prominent in this field, for which the self-explanatory term 'osteocytic osteolysis' has come to be accepted.

The evidence for the occurrence of osteocytic osteolysis comes from observations of three kinds; morphological, cytochemical, and biochemical. The chief points are as follows:

(a) MORPHOLOGICAL

A basic fact is that some osteocytes normally look larger than others. It has been claimed that such enlarged osteocytes increase in number and size following administration of parathyroid extract, which of course stimulates bone resorption. Similar changes have been reported in the medullary bone of the laying hen (Taylor & Belanger, 1969). These osteocytes can be seen with a relatively unsophisticated technique such as the light microscopy of routine sections, but have been much studied by more complicated methods, like microradiography, which reveal an apparent perilacunar demineralization around them. When microradiographs of decalcified bone sections are made with alpha radiation emitted from

[146]

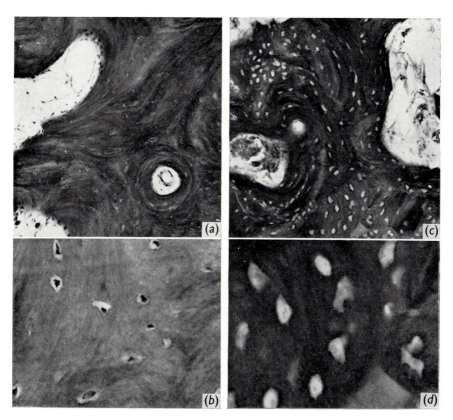

Fig. 75. Tissue from region of knee joint from patient suffering from osteochondritis dissecans.

(*a*) Apparently normal live bone, magnification, × 120. Osteocyte lacunae are relatively small and inconspicuous.

(*b*) As above; the osteocytes appear as dark dots within the lacunae. Magnification, × 480.

(*c*) Dead bone. Same magnification as (*a*). Lacunae are more distinct because they are bigger and empty.

(*d*) Part of (*c*); magnification as in (*b*). Lacunae are bigger than in (*b*), and lack osteocytes.

a polonium source, the organic matrix appears to be reduced around some lacunae (Belanger *et al.* 1963, 1967). Electron microscope studies (Baud, 1966, 1968; Belanger *et al.* 1966; Jande & Belanger, 1969), show some lacunae to be surrounded by matrix which, in places, is incompletely calcified.

A persuasive picture of apparent enlargement of lacunae is often to be seen in 'dead' bone. In the example illustrated (Fig. 75), the specimen came from a patient with osteochondritis dissecans and provided samples of living and dead bone side by side with each other. The lacunae of the dead bone look very much enlarged, but this is almost certainly an artefactual post-mortem change.

Recent work by Dame Honor Fell and her associates provides another example of how, under abnormal conditions, bone cells can be persuaded to do unusual things. When embryonic bone is grown *in vitro* in the presence of sucrose, the periosteal cells, osteoblasts and osteocytes become loaded with vacuoles (Fell & Dingle, 1969). Osteocyte lacunae are enlarged; the matrix looks different and stains differently from normal. Electron microscope studies suggest that the exposure to sucrose stimulates an increased production of lysosomal enzymes and that some are released into the surrounding matrix. The latter is degraded; osteoblasts and fibroblasts then phagocytose matrix constituents. Though not specifically mentioned, osteocytes presumably do the same (Glauert, Fell & Dingle, 1969). To what extent, if any, this cellular response to an abnormal environment operates in normal everyday life is uncertain.

How should one interpret observations of these kinds? Do they mean that the osteocytes and lacunae are swelling, and that bone is disappearing? Or do they mean the osteocytes and lacunae are shrinking and bone is being laid down? The only objective proof would, of course, be impossible to obtain; that is, a cine record instead of a still picture. One must keep an open mind on this question, because it is known that the sizes of osteocyte lacunae normally vary considerably. In newly formed bone, the lacunae are often relatively large, whereas in lamellar bone, the more mature matrix is usually populated by smaller lacunae which, however, vary in size. Again, electron micrographs do show considerable, apparently normal variations in the mineral density around lacunae both in newly formed bone (e.g. Figs. 46–50) and in adult bone (Cooper *et al.* 1966); the lacunar wall also varies considerably in the smoothness of its contour, and in the closeness of the fit of its contained osteocyte.

More objectively, electron microscopy of osteoclasts has demonstrated that these cells lack the ruffled border of osteocytes and there are no signs of the endocytosis of mineral crystals or collagen fragments. If osteocytes do resorb bone, it must be in a way completely different from that of the osteoclast.

(*b*) CYTOCHEMICAL

After administration of parathyroid extract, staining with the PAS method sometimes shows differences in the matrix around some osteocytes, in that it assumes a deeper intensity and colour of staining. This is thought to reflect a relative depolymerization of the ground substance.

A fairly extensive list of enzymes which could be involved in osteolysis have been demonstrated in osteocytes by cytochemical methods; many of these are characteristic for the spectrum of lysosome enzymes. A full list (and references) is given by Vaes (1969). There have been several claims that lysosomes can themselves be visualized in osteocytes (Baud, 1966, 1968; Belanger *et al.* 1966), but this is on morphological rather than cyto-chemical grounds.

One particular cytochemical result should be mentioned for its technical ingenuity. Bone sections were placed on the surface of blackened photo-graphic emulsion and incubated under controlled pH conditions. After an hour, the gelatin beneath had been digested, and so a transparent 'hole' was apparent (Belanger & Migikovsky, 1963). This 'protease' (?gelatinase) was more effective at neutral than acid pH. To keep matters in perspective, it should be said that there is no evidence that the 'gelatinase' revealed by this method, employed also many years ago to demonstrate pepsin (Dorris, 1935), has any effect on bone matrix.

Whilst cytochemistry tells us that osteocytes have a variety of hydro-lytic enzymes in their cytoplasm, it is interesting and possibly significant that, according to the list given by Vaes (1969), there are only two differ-ences between their enzyme endowment and that of osteoblasts. The latter have phosphoamidase, but lack protease. However, that osteoblasts can certainly exercise a proteolytic effect is shown by the way they liquefy a plasma clot in tissue culture. The question then is whether and why osteocytes use their enzymes to degrade surrounding bone when osteo-blasts, with a similar enzyme battery, act the opposite way. To this ques-tion, cytochemistry gives no answer.

(c) BIOCHEMICAL

Biochemical methods in general offer the hope of quantitative work and the possibility of identifying constituents undetectable by cytochemical technique. However, the only way in which bone tissue can be prepared for biochemical study is to homogenize it. It is not feasible to harvest the different kinds of cells separately with any confidence. Results with homo-genates cannot tell us about the attributes of the separate cell races, the osteocytes, osteoblasts, osteoclasts, myeloid and fat cells. They inform us about the sum of the properties of these cells and of the intercellular matrix. Thus Vaes (1965) worked with baby rat skull in which he found and assayed various hydrolytic enzymes. Other workers have found col-lagenases. Nichols (1970) refers to results with 'preparations rich in osteocytes' but, here, too, the identification of the cells cannot be made in a really objective and reliable manner. For these reasons, unfortunately, biochemical work has not so far been able to provide much infor-mation about the existence or mechanism of osteocytic osteolysis.

In summary, it is fair to say that the evidence for osteolytic osteolysis, though persuasive, is still not absolutely conclusive. Several difficulties remain, and one, at least, has so far hardly received attention, namely, the actual physical transportation of the products of osteolysis away from the scene. Osteolytic osteolysis is said to affect even the most mature, deep-seated lacunae in the very depths of bone trabeculae. A pathway to the outside world theoretically exists in the inter-connecting canaliculi and lacunae, at various depths in the bone, which leads ultimately, one supposes, to the nearest Haversian canal. However, this only applies to the Haversian lamellae. Osteocytes within interstitial lamellae presumably cannot communicate with the exterior in this way for, as pointed out on p. 30, their canaliculi do not as a rule make outside connections. Even for the Haversian osteocyte canaliculi and lacunae, although a track through the bone may exist, its direction, from endosteal surface onwards, is relatively uncharted, and travel along it must surely be very slow.

Nevertheless, even if its reality is doubtful, the concept of pericanalicular and perilacunar de- and perhaps remineralization is extremely attractive. It has been applied to the eel as well as the mammal (Lopez, 1970). The surface area provided by the body's canaliculi and lacunae must be staggeringly large; fluctuations in mineral concentrations at this surface sufficient to influence blood calcium levels would probably be so minute as to be quite undetectable by any method currently available. This kind of traffic, for all we know, may be an important everyday biological mechanism. Recent reports (Nichols, 1970; Talmage, 1967) propose that in the ordinary way the serum calcium level is maintained via the agency of the osteocytes; mineral, it is suggested, leaves the walls of the canaliculi and/or lacunae, travels the transosteocytic lacunar and canalicular pathways and gains access to the blood. Moss (1963) has claimed that whereas fish with cellular bone can withdraw calcium from bone to mineralize fracture callus, other species possessing acellular bone cannot do so.

2. MAST CELLS AND BONE RESORPTION

It has been known for many years that mast cells can be demonstrated in the bone marrow and sometimes on endosteal bone surfaces. The possibility that they play a role in resorption depends upon circumstantial rather than direct evidence, of the following sort. First, there have been several reports that the number of mast cells present in the bones of patients suffering from osteoporosis is increased (Frame & Nixon, 1968). In this condition the mass of bone tissue is greatly diminished, probably as the end result of a rather slow resorptive process. Second, the long-term administration of heparin (of which substance the mast cell is believed to be the natural source) to patients gives rise to osteoporosis (Griffith *et al.* 1965); whilst, in tissue culture experiments, heparin acts as a cofactor in stimulating bone resorption by osteoclasts (Goldhaber, 1965). Third, administration of relatively small doses of parathyroid extract (a potent stimulus to bone resorption) for long periods to weanling rats, or a calcium deficient diet, produce an increase in the bone mast cells (Rockoff & Armstrong, 1970). Fourth, Severson (1969) discovered that mast cells proliferate in areas of bone resorption produced experimentally and proposed that they are involved in the release of hydrolytic enzymes. Finally, Lindholm and associates (1969) demonstrated the occurrence of mast cells in healing fractures, studied the effects of various experimental procedures upon the numbers present, and concluded that the mast cell's granules play some fundamental role.

It would be fair to say that at the present time there is no convincing evidence to show that mast cells take a direct part in the removal of bone, which can certainly be carried out by osteoclasts in the absence of mast cells. On the other hand, there does seem to be some kind of intriguing association between mast cells and bone resorption.

15

UNKNOWN FACTORS IN
BONE RESORPTION

In many instances, resorption of bone can be referred to known causes, such as parathyroid hormone, calcium deficiency, or local pressure. Sometimes, however, the aetiology of the resorption remains a mystery; and sometimes, conversely, there is an unexplained failure of resorption in circumstances where it would normally occur. A few instances of these intriguing situations now follow.

1. DISUSE ATROPHY

The bones of a limb immobilized through muscle paralysis, or by being held still in a plaster cast, soon undergo quite profound, retrograde changes which are described as disuse atrophy or osteoporosis. Although the external shape and size of the bones themselves show no appreciable alterations, their density is much reduced; the quantity of bone tissue may be greatly diminished so that trabeculae become thinner, and large resorption cavities may be present. The X-ray shadow of such bones is characteristic (Fig. 76). They appear less radiopaque, and the pattern of the trabeculae themselves sometimes becomes apparent against the background of overall reduced density. If the plaster is removed within normal time limits and the limb soon comes back into use, the bones return quickly to normal.

The explanation of these rather striking changes is an intriguing problem. The fact that the rest of the bones of the skeleton remain normal shows that a local factor must be involved. At one time it was widely believed that the cause was to be found in disturbance of the blood flow through the bones, the lack of muscular contractions leading to a condition of venous engorgement. This, in its turn, has been regarded as the stimulus for the osteoclastic activity responsible for the loss of bone tissue (Trueta, 1963). Matters are not so simple as this, however. Under normal conditions, the turnover of bone is under endocrine control; the role of the parathyroid gland has been described in the preceding section and the effects of its hormone are, of course, widespread and certainly bilateral. Experiments (Burkhart & Jowsey, 1967) have shown, however, that the unilateral and

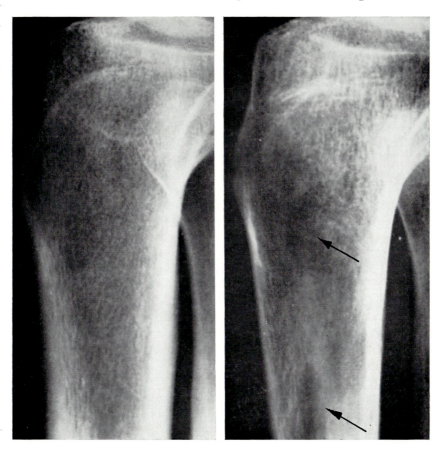

Fig. 76. X-rays of upper end of human tibiae. Left, normal; right, from patient unable to work for 10 years because of muscular dystrophy. There is an overall loss of density and patches of rarefaction (arrows) are present.

strictly localized resorption induced by plaster immobilization is parathyroid- and thyroid-dependent. This was demonstrated by experiments in which limbs (dog) were immobilized in plaster. In control animals, there soon followed a marked disuse atrophy, which was studied by microradiography. However, in animals whose thyroids and parathyroids were removed before limb immobilization, no bone changes occurred. This leaves two possibilities open; either the localized changes of disuse atrophy, thus shown to be endocrine dependent, involve the production of a local factor which intensifies a hormonal stimulus so weak as to have no effect on the non-immobilized contralateral bone; or there may be some kind of local primary, underlying damage to or change in bone tissue which makes it unusually susceptible to the effects of the hormone. In this latter context it might be that the nervous system plays a role. The changes in the bones and calcium metabolism which occur in astronauts following prolonged weightlessness in space presumably provide another example of disuse atrophy.

2. DISAPPEARING BONE DISEASE

Another example of abnormal resorption is found in the rare and bizarre condition of 'disappearing bone' or 'massive osteolysis', which is encountered in children and young adults.

The characteristic feature is a bone resorption, often more or less painless, of extraordinary intensity and extent. Bones of the arms, the ribs, the pelvis and legs may be affected. Within a few months or years most of an ischium or ilium may simply disappear. Edeiken & Hoder (1967) illustrate a remarkable case in which practically the whole ilium melted away. In another case (Fiore & Smyth, 1960), in which the pelvis was affected, most of the superior and inferior pubic rami had disappeared on one side; following radiotherapy, within six months the lost bone was restored, remarkable enough on its own but, even more striking, the shape and X-ray appearance was indistinguishable from normal. The disease, however, eventually recurred and the patient died.

The cause of this extreme resorption is not known for certain. The parathyroids are apparently normal. It is usually thought that the bone changes are secondary to invasion by a vascular tumour (Halliday *et al.* 1964), but on the other hand, the presence of haemangioma-like tissue in the ruins of the bone might reflect hypervascularity resulting from attempts at healing. As well as the extraordinary severity of the disease, another fascinating problem is that of localization. Why a particular site should be affected remains an enigma.

3. 'GREY LETHAL' MUTANT IN MICE

In this strain, a proportion of the offspring show, within a few days of birth, an abnormality of the colour and texture of the coat. Later on they develop bony abnormalities and their teeth do not erupt; they die within a few weeks.

The condition was originally reported by Grüneberg (1936). The skeletal defect is, essentially, a patchy failure of bone resorption, so that vascular and nervous foramina in bones do not enlarge to accommodate the increasing girth of their soft tissue contents; teeth are retained within the maxilla and mandible because their eruption is hindered by the failure of the overlying bone to resorb normally.

There could be three possible reasons why the bone of grey lethals fails to resorb normally. First, a defect in the bone tissue such that in spite of the presence of normal osteoclasts and parathyroid hormone it remains, as it were, insoluble; second, some inadequacy of the osteoclasts which are incapable of resorbing normal bone; or third, some parathyroid abnormality which results in the failure of some vital message which normally causes osteoclasts to erode bone. Barnicot (1941) has worked on this problem, grafting normal bone into grey lethal hosts and vice versa; injecting grey lethals with parathyroid extract (1945); and comparing the number and size of the osteoclasts on the skull bones of normal and grey lethal animals (1947). The evidence he has obtained suggests that the basic defect resides in the osteoclasts. Exactly what their shortcomings are remains to be discovered.

4. CLEIDO-CRANIAL DYSOSTOSIS IN MAN

This affords another example of a congenital condition in which, whilst a large part of the skeleton may be normal, there are also unexplained aberrations of both bone resorption and deposition. In this dysostosis some (or even many) of the bones normally arising by intramembranous as opposed to endochondral ossification, for example, skull bones, clavicles, fail to form at all. Yet there is also evidence of failure of resorption. This is manifest from the non-eruption and bizarre shape of some of the teeth which are ridged longitudinally and twisted in a manner reminiscent of grey lethals.

The existence of conditions such as this (and there are many other equally strange inherited bone abnormalities), though intrinsically most interesting, has a wider significance in relation to our concepts of bone formation and resorption. It underlines the fact that though hormonal factors like parathormone and calcitonin may exercise a general overall influence upon osteoclasts and osteoblasts there is evidently a wide degree of local autonomy. It reveals also how the differentiation of normal osteoblasts must depend on local factors. Why they should fail to appear for the construction of a clavicle yet be present normally for humerus is likely to be an insoluble mystery.

SECTION 4

OSTEOGENIC STIMULI

16

OSTEOGENIC STIMULI

As with bone resorption, the deposition of bone occurs both as a normal event and also in a variety of pathological conditions. It, too, may be provoked experimentally. The principal histological events in the production of bone, wherever and whenever it occurs, are the appearance on the scene of osteoblasts, and the inception by them of osteogenesis. Thus, factors which provoke the differentiation and activity of osteoblasts may be considered as osteogenic stimuli. They fall into two general categories; first, those that apparently emanate from a discernible, localized source, and, second, those of a more general, blood-borne nature.

Good examples of local factors are provided by the healing of fractures; by the response of bone surrounding teeth influenced by orthodontic appliances; and by the occurrence of bone in unusual sites outside the skeleton as happens in a variety of pathological conditions, and as may be provoked by deliberate experimental procedures.

The healing of a fracture, of which good brief accounts are given by McLean & Urist (1968) and Pritchard (1964), is accomplished through a number of consecutive stages. At the beginning, there is the removal of blood clot and other debris by mobile phagocytic cells and the creation of a highly cellular, richly vascular connective tissue. Meantime, the broken ends of the bone undergo changes. At the broken edges and some little distance within the bone the osteocytes are damaged and perish. This 'dead' bone is, at least in part, removed by osteoclasts. Meantime, large quantities of fibrocartilage appear in the new young connective tissue; it bridges the gap between the broken ends and helps to immobilize them. This bulky fibrocartilagenous callus is, however, but a temporary expedient. A little later, osteoblasts appear, probably originating from pre-existing osteogenic cells of the periosteum, and lay down trabeculae of woven bone. These, in the case of a long fracture, originate some little distance back from the broken ends; the trabeculae grow across the gap, more or less beneath the fibrocartilage, to unite the fractured ends. Extensive remodelling of the newly-formed as well as the original bone near the broken ends takes place; the fibrocartilage is removed, swelling disappears, and the bone contour slowly returns to normal.

Another example of the operation of a local factor is provided by the reaction of the bone around teeth which are subjected to orthodontic procedures. The latter are undertaken to move teeth about and this is effected by gentle pressure, usually provided by metallic spring devices. When a tooth is moved in this way, gentle force is exerted on the bone of the tooth socket via the parodontal ligament, a strong band of collagenous fibre connecting tooth to jaw bone. It is found that, where the bone is subjected to compression, resorption by osteoclasts is stimulated; on the opposite side, subjected to pull, osteoblastic activity and the laying down of new bone takes place.

A third example of local factors in the stimulation of osteogenesis is seen in the appearance of bone in unusual places which occurs in some pathological conditions. For instance, in man, a progressive ossification of skeletal muscle occurs in the condition known as myositis ossificans; bone may develop in the wall of the aorta of the horse infected with strongyloides and has been reported in the stomach mucosa following subtotal gastrectomy in the rat (Ackerman, 1968).

The naturally occurring conditions under which local factors seem to stimulate osteogenesis do not afford promising material in which to identify osteogenic stimuli. In the case of fracture repair, we have to deal with an enormously complicated series of changes which involve not only the synthesis but also the breakdown and removal of complex intercellular products. It seems practically certain that this stepwise sequence of events is triggered off by a whole series of consecutive stimuli of local origin. How to unravel the effects of these is a baffling problem. Myositis ossificans, though a simpler situation, is hardly amenable to experimental analysis; but, at least in principle, although little seems in practice to have been accomplished, one can imagine how experiments might shed light on the mode of action of strongyloides in inducing the formation of bone.

In effect, our knowledge about osteogenic stimuli stems very largely from experiments with what is generally called bone induction; that is, the production of bone by cells of the host animal around an experimental implant. Many different kinds of substances have been used for such implantation experiments. They may be divided into two groups; the first comprises various living tissues, and the second, non-living materials. Results of implantations will now be described.

17

BONE INDUCTION BY
LIVING IMPLANTS

A wide variety of living tissues can emit an osteogenic stimulus when transplanted intramuscularly, as the following account shows.

1. INDUCTION OF BONE BY
TRANSITIONAL EPITHELIUM

A bizarre and unexpected property of the epithelium lining the urinary bladder and ureters is its ability, on transplantation to a new site, to induce the appearance of bone in its immediate vicinity. Some of the classical work on this strange phenomenon was carried out by the American surgeon, Charles Huggins, on dogs (1931). He showed that following the successful transplantation of small pieces of bladder mucosa, cysts developed which contained a lumen lined by uroepithelium consisting partly of the original and partly of newly-proliferated transitional epithelium. The cyst wall, of connective tissue, contained newly-formed bone close to the epithelium.

He found that whilst the epithelium induced bone especially successfully when transplanted to the muscle of the anterior abdominal wall, negative results were consistently obtained with the spleen, liver and kidney parenchyma even though the epithelium survived there. Similarly, the wall of the bladder itself proved a refractory site. Huggins demonstrated that the source of the osteogenic stimulus resided in the epithelial cell layer, rather than the subjacent mucosa, by implanting bladder mucosa into the abdominal muscles on one side, and the remaining elements of the bladder wall on the other. The former consistently gave positive and the latter negative results. He also gently scraped the bladder surface with a scalpel and transferred the detached epithelial cells to a muscle pocket; new bone was induced. This further emphasized that the newly formed bone is found in association with the epithelial elements of the bladder. Epithelium from the gall-bladder was also shown to be osteoinductive although very much more feebly so than uroepithelium. Gastric, jejunal and prostatic epithelium gave negative results as did implants of adrenal.

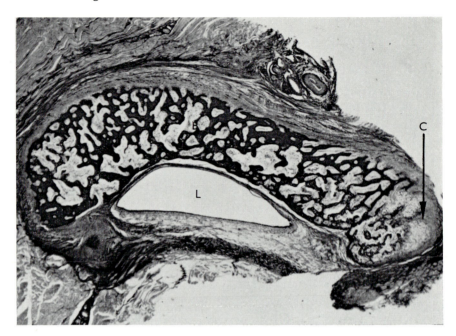

Fig. 77. Section showing bone induced by transplanted transitional epithelium in the pig. A cyst lumen (L) is lined by the epithelium though it cannot be identified at this magnification. There is a mass of newly-formed bone (B) nearby, and of cartilage (C) undergoing endochondral ossification. Magnification, × 16.

From these experimental findings, Huggins deduced that living uroepithelial cells secrete a substance which acts upon the young, reactive connective tissue. In the case of skeletal muscle, this causes the differentiation of osteoblasts from the fibroblast-like cells, but, with spleen and urinary bladder wall, the fibroblasts either do not heed the message or are unable to respond to it.

Huggins' results in the dog have since been tested out many times in other species, in most of which a positive result is obtained, i.e. in man, cat, mouse, pig (Fig. 77), rat, hamster, guinea-pig (Fig. 78); but in a few, i.e. rabbit, goat, although the epithelium may survive and appear perfectly healthy, bone does not form in the cyst wall. The reason for these species differences is quite obscure. Typical histological findings are illustrated in Figs. 77 and 78. The very close proximity of the new bone to the epithelium is a striking and constant finding.

Uroepithelial transplants between different donors and recipients of the same species ('allogenic') survive in a proportion of cases long enough for the osteogenic stimulus to be effective, and new bone forms. Soon afterwards the immune defence mechanisms of the host supervene and the

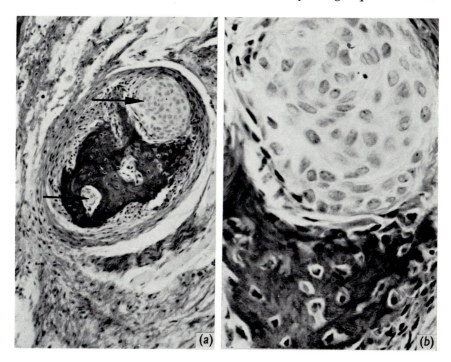

Fig. 78. Osteoinductive effect of uroepithelium in the guinea-pig. Magnification, × 120.

(*a*) Small aggregates of transitional epithelium had been implanted intramuscularly fourteen days previously. A larger (upper arrow) and a smaller (lower arrow) group of epithelial cells are present; new bone (B) has formed nearby them.

(*b*) Part of upper epithelial cell next at increased magnification (× 480), to show the close relationship of the new bone to epithelial cells.

transplanted cells die. The bone and osteoblasts, however, survive for a time, thus proving that they originate from host, not donor. After a further interval the osteoblasts disappear and the bone is gradually resorbed by osteoclasts. This has been taken to indicate that a continuous flow of osteogenic stimulus is needed not only for the appearance of osteoblasts but also for their persistence (Friedenstein *et al.* 1967).

In more recent times, work on the osteogenic properties of transitional epithelium has been carried forward with techniques employing cell suspensions and millipore diffusion chambers. With the former, uroepithelium is subjected to treatments evolved for the dissociation of tissues and organs into their constituent cells for use as suspensions of single cells or small groups of cells (Moscona, 1952). Such treatment usually entails short incubation of fresh tissue in dilute trypsin–pancreatin solutions in calcium- and magnesium-free saline, followed by washing and gentle physical pressure which produces a suspension of separated epithelial cells and cell clumps. When injected into guinea-pigs, suspended uroepithelial cells form solid cords or nests rather than hollow cysts. Bone forms around them (Moskalewski, 1963; Ostrowski *et al.* 1967). These results rule out the possibility raised by some workers that the fluid usually found in bone-bearing cysts is the source of the osteoinductive factor. They also demonstrate the need for living epithelium, because negative results followed the injection of killed cells. Uroepithelium is, of course, stratified, and the various layers are characterized by cells of varying fine structure. Those on the surface, for instance, are very large indeed and in the electron microscope are seen to possess curious angular vesicles. At present it is unknown whether the osteogenic factor derives from any particular layer or type of epithelial cell, but cell separation and suspension techniques might be developed to settle the question.

Successful cross-species bone induction by uroepithelial transplants has been obtained recently (Włodarski, 1970). The host animals were injected with cortisone to suppress their immune responses; the foreign epithelium then survived and bone induction occurred.

Millipore, a form of cellulose acetate, is produced commercially, as sheets, of various thickness, penetrated by pores of known size. In biological work, a pore diameter of about half a micron is usually chosen. A diffusion chamber can be constructed by taking a ring of Perspex and cementing sheets of millipore across both sides to make a hollow, drumlike object. For work with living tissue, millipore is first cemented to one side of the Perspex ring; then the tissue is introduced, and finally the millipore sheet is cemented in place across the other side. Such diffusion chambers, known as Algire chambers after the inventor (Algire, 1957), offer some strong theoretical advantages for work upon osteoinduction. When a tissue-bearing chamber is implanted in an animal, the cells within often survive and flourish, because substances can diffuse into and out of the chamber through the pores in its wall. Further, the cells within the chamber are given complete shelter from attack by host immune response defensive cells, which cannot penetrate the pores. Thus, allogenic cells (that is, grafted or transferred from one individual to another of the same species) or even xenogenic (from one individual to another from a different species) may not only survive, but proliferate and differentiate within the chamber.

These theoretical advantages cannot always be guaranteed; cracks or small perforations in the filter wall or detachments of filter from the Perspex ring may be impossible to detect (e.g. Post *et al.* 1966). Of course, if capillaries are seen within the chamber a defect must certainly have been present, but it is never possible to rule out absolutely that the wall may have been leaky to cells, even if blood vessels are missing. However, subject to this very important reservation, some interesting and basic information about osteoinduction has been obtained with millipore chambers. To begin with, there is the question as to whether bone will form around diffusion chambers containing transitional epithelium. Unfortunately, the answers are contradictory. According to Friedenstein (1962, 1968) it does; on the other hand, according to Ostrowski *et al.* (1967) it does not. The validity of experiments along these lines is also complicated by an even more fundamental disagreement, namely upon results of implanting millipore material on its own, or as empty chambers. For the most part, opinion is that millipore itself does not emit an osteoinductive stimulus, but there are isolated reports to the contrary.

Diffusion chambers have been used in another way, which does not entirely depend on the complete integrity of the filter membranes. Friedenstein (1968) mixed suspensions of dissociated guinea-pig uroepithelial cells with suspensions of cells prepared from other tissues, placed a volume of the mixture within a chamber and implanted it intraperitoneally or subcutaneously in guinea-pigs. The cells survived. When peritoneal macrophages, white blood or spleen cells had been added, bone formed within the diffusion chamber; but with connective tissue cells from the tunica propria of the bladder, or from subcutaneous connective tissue, there was no bone. It is interesting that these results with trypsin-isolated cells seem to conflict with Huggins' original findings with the intact spleen.

At the present time, the source of the uroepithelial osteoinductive factor remains to be identified, as does also its mode of action. The fact that even xenogenic transitional epithelial cells, maintained in cortisone-conditioned mice (Włodarski, 1970), can induce bone seems to point to some strong and consistent effect.

2. INDUCTION OF BONE BY EPITHELIAL CELL LINES GROWN IN TISSUE CULTURE

As strange as the osteoinductive effect of urinary tract epithelium is that of human amnion tissue-culture cell lines. The first reports that bone appeared around suspensions of such cells when injected into mice, whose cellular immune response had been inhibited by cortisone, appeared as far back as 1964 (Anderson *et al.*), when it was shown that cells of the 'FL' line had this property. This has since been confirmed (Anderson, 1967) and it has also been shown that other 'lines' of human amnion cells *in vitro* have similar properties. Indeed, it has now been shown that several lines of epithelial cells can induce bone and it may even prove to be a general property of all established epithelial cell lines (Włodarski, 1970).

The host tissue responses to tissue culture epithelial cell lines and transitional epithelium differ slightly. In the latter case, as described above, osteoblasts appear and bone is generally, but not always (see Fig. 77), directly deposited. With the former, cartilage is normally produced in quantity and it seems likely that most of the bone is formed by endochondral ossification of the cartilage. This difference must have biological significance but what this may be is enigmatical at present.

All attempts to isolate the osteogenic stimulus have failed, as with transitional epithelium. The cytoplasmic organization of transplanted epithelial cells surviving in cortisone-conditioned animals is not that of secretory cells. It lacks endoplasmic reticulum and the Golgi apparatus is poorly developed. Electron microscope appearances do not suggest that a protein or mucopolysaccharide product is synthesized.

The osteogenic effect of epithelial implants is generally explained as an effect on connective tissue cells which respond by turning into or giving rise to osteoblasts. In a way, it is rather surprising that other mesenchyme-derived tissues, such as muscle, do not also form. This suggests that whatever the stimulus is, it must possess a rather high degree of specificity. It has, of course, been known for many years that epithelium has other effects on underlying connective tissue. For example, regeneration of transected anuran limbs is influenced by the epidermal epithelium; if a cuff of skin is removed proximal to the cut surface, regeneration fails as it also does if gut is substituted for skin epithelium. It has also been shown that the healing of skin wounds *in vitro* is greatly influenced by an effect of the newly formed epithelial sheet on the orientation of underlying collagen fibres (Bentley, 1936). A particularly dramatic effect of skin epithelium on wound healing is seen in the horse. In this species, cuts in the lower extremities heal very slowly if at all. There is a massive overgrowth of vascular 'granulation' tissue which sprouts forth and does not permit the natural growth of epithelium across its surface. However, if small pieces of skin are grafted or implanted into the granulation tissue, the overgrowth of the latter ceases and the wound at last heals. Such effects as these seem to fall within the general category of relationships now known as epitheliomesenchymal reactions.

Disappointing for those interested in bone induction is the fact that much more is known about effects of mesenchyme on epithelium than vice versa. However, the former may elucidate the latter, and so a few of the more striking examples of epithelial responses to mesenchyme will next be described.

Most of the work on this fascinating topic has been carried out with embryonic tissues. The principle is to select a particular type of epithelium; either exocrine glandular, as, for example, the embryonic pancreas or salivary gland; or endocrine, as in the thyroid; or as an epithelial sheet, for instance, the skin. The selected epithelium is then used as an indicator against which can be tested certain effects of mesenchyme. The indications are the light- and electron-microscopic appearance of the epithelial cells, and their ability to undergo normal morphogenesis in the absence of their normal mesenchymal substratum or in the presence of mesenchyme obtained from other sites. Such mesenchyme samples are obtained, and epithelia are separated, by techniques for organ dissociation employing tryptic digestion, etc. The two kinds of tissues, epithelial and mesenchymal, are confronted with each other in tissue culture or as grafts on the chorio-allantoic membrane.

It transpires that exocrine glandular epithelium is dependent on mesenchyme for its proper development, and that each may influence the other. For instance, to quote from Grobstein, (1968):

From these studies certain general properties have emerged. First, epithelium is dependent upon mesenchymal association for continued development. Second, the epithelium is specific; i.e. it has some information in advance of the interaction as to what kind it is. Pancreatic epithelium across from salivary mesenchyme differentiates as pancreas; salivary epithelium across from the same mesenchyme differentiates as salivary. Third, not only does the epithelium have information, but the mesenchyme does as well. One can get different reactions depending on the kind of mesenchyme present. Salivary mesenchyme with salivary epithelium yields characteristic epithelial morphogenesis, but jaw mesenchyme with salivary epithelium gives a compact epithelial mass with no morphogenesis. Clearly, some difference exists between salivary mesenchyme and jaw mesenchyme. Fourth, not only are epithelium and mesenchyme specific in properties, but there is also specificity of varying degree in the interaction itself. Some epithelia, like pancreatic epithelium, are able to continue their characteristic morphogenesis in the presence of mesenchyme which is not their own – in fact, pancreatic epithelium appears to continue its development in the presence of any mesenchyme so far tested. Salivary epithelium, on the other hand, is quite specific in its requirement: it undergoes characteristic morphogenesis only in the presence of its own kind of mesenchyme. Other epithelia are intermediate between salivary and pancreas, reacting with mesenchyme other than their own but not with all mesenchyme. These variations in the specificity of interaction, as well as the other general properties referred to, must be taken into account in considerations of mechanism.

It has been shown (Hilfer, 1968) that thyroid cells retain their characteristic ultrastructure only in the presence of constituents from the connective tissue capsule, an effect probably mediated by capsular fibroblasts. In the case of skin epithelium, denuded of its own supportive mesenchyme, appearances varied widely according to whether it was planted in limb, gizzard or heart mesenchyme (McLoughlin, 1968).

Experiments have established that such effects of mesenchyme on epithelium can pass through millipore filters, ruling out the possibility that cell-to-cell contacts are essential for the transmission and receipt of the morphogenic stimulus. Indeed, work with radiolabelled compounds has shown that passage of actual substances by diffusion is involved in transfilter effects (Grobstein, 1968; Bernfield, 1970). What these may be, and how they operate, remain to be discovered, though the ubiquitous collagen molecule may be implicated.

3. INDUCTION BY BONE ITSELF

It has been common knowledge for a very long time that newly formed bone arises in association with certain types of living auto- and homografts of bone. In the latter instance, new bone populated by healthy-looking, active osteoblasts and osteocytes appears within a few weeks near or around the graft, and may persist for many months or even years. This indicates very strongly that the surviving osteogenic cells must be of host not graft origin, for otherwise they would have perished soon after implantation through attack by the host's immune responses; this observation affords one good line of evidence that even bone matrix itself can emit an osteogenic signal. The same inductive mechanism probably operates in the case of new bone forming around small pieces of dead bone (Fig. 79).

Other evidence comes from work with millipore diffusion chambers. The original observation was published in 1961 by Goldhaber. Working with mice, he implanted pieces of allogenic neonatal skull bone in chambers. The bone within survived and new bone was laid down on the outside of the adjoining filter wall even in hosts whose immunity against donor tissue had been heightened in advance. This result has been confirmed by several others. The experiment apparently does not succeed if long bone material is used (Post *et al.* 1966), but the significance of this finding is difficult to judge.

Work has also been carried out, in mice, with diffusion chambers containing implants of an osteogenic sarcoma (Friedman *et al.* 1968). The osteoblasts of the latter were distinguishable from normal osteoblasts in electron micrographs; virus particles were seen in association with the former, but not, of course, with the latter. Bone was seen outside the filter in association with osteogenic sarcomatous bone inside; the induced bone was populated by normal-looking, rather than sarcomatous osteoblasts; virus particles were absent.

Some further observations were made which are very important in the interpretation of the so-called 'transfilter' induction of bone. Induced bone is always seen to arise on the outside surface of areas of millipore which bears living bone on its opposite (inner) surface. Van Gieson staining of the filter material, separating implant from induced bone, revealed the presence of fine fibrils. Electron microscopy showed that these were collagenous, and that they traversed the whole thickness of the filter wall; they were coated with a fine filamentous material. Cell processes did not penetrate the thickness of the wall so that direct cell-to-cell contacts could be excluded (Post *et al.* 1966). In addition, collagen fibrils were absent from the filter wall in areas where induced bone was absent. These observations have been taken to suggest that the osteogenic signal emanating from the outer surface of the filter comes from either the collagen fibrils or the associated filamentous material.

4. BONE MARROW

Osteogenic properties are seemingly possessed by bone marrow, as well as by epithelia and bone itself. One of the early reports on bone induction by marrow referred to the development of bone around marrow implants in the anterior chamber of the eye (Scowen, 1940, 1941, 1942). In another series of experiments, bone induction occurred with 100 per cent success following autografts and 80 per cent with allogenic grafts (Danis, 1957); other experiments, not quite as successful, were described by Levander (1964). There seems to have been little experimental work to investigate the nature of the stimulus emitted by myeloid tissue, though this tissue offers several interesting possibilities.

5. INDUCTION OF CARTILAGE

A further example of tissue interaction is relevant to the induction of bone. However, in this instance the effect is not one between epithelial and mesenchymal components leading to the appearance of bone, but is concerned with the embryonic spinal cord acting to induce the appearance of cartilage.

Experiments *in vivo* and *in vitro* have now established (Lash, 1968) certain facts about the relationship between these two tissues. If notochord and spinal cord are extirpated from a developing embryo, the cartilage expected to develop from adjacent somites fails to do so. If pieces of notochord or cord are implanted into a part of the somite which would normally make muscle, it is induced to construct cartilage instead. *In vitro*, somites produce fibroblasts, pigment cells and immature muscle elements; but if a piece of notochord is present, cultures produce masses of cartilage. Even more dramatic is the fact that cell-free extracts of spinal cord or notochord will do the same thing. The stimulus emanating from the cord or notochord is capable of passing through a millipore filter; its effect on somite cells requires only a few hours of contact, but cartilage does not appear until four days later.

These findings, until recently, were widely interpreted as demonstrating that some special substance secreted by cord or notochord instructs undifferentiated somite cells to synthesize the collagen and the sulphated mucopolysaccharides of cartilage matrix. However, matters are not this simple, for it has now become apparent that these same somite cells will form cartilage *in vitro* spontaneously under certain conditions of culture, even if not profusely. This implies that the reactive somite cells, which had hitherto been regarded as indifferent or 'undifferentiated', must, in fact, already be partly differentiated at the time the inducer acts.

It is very difficult to relate cartilage induction by spinal cord to that of bone by, say, transitional epithelial cells. In some ways the latter seems a more specific effect. For instance, when epithelial cells are maintained together with peritoneal macrophages in millipore diffusion chambers, bone is said to form within (Friedenstein, 1968). Yet peritoneal macrophages never form bone on their own.

18

BONE INDUCTION BY
NON-LIVING IMPLANTS

Quite a range of non-living materials, some of them most unexpected, emit an osteogenic stimulus when implanted in animals. They can be either of biological or non-biological origin.

1. BIOLOGICAL SUBSTANCES

Certain tissues when freeze-dried (lyophilized) and implanted in a suitable site can induce the formation of bone, whilst others can not. The former include placenta and urinary bladder with which the percentage rate of successful induction is, however, relatively very low and the yield of bone scanty (Urist *et al.* 1967, 1968). In contrast, when certain hard tissues are decalcified in a prescribed way and the remaining collagen-soft tissue framework is lyophilized and implanted, bone is induced to form in a very high proportion of experiments. Tissues giving especially good results are bone and dentine (Figs. 79, 80 & 81). Several aspects of this intriguing phenomenon have been investigated by Urist and his colleagues.

First, there is the question of how methods of preparing the bone samples for implantation affect their osteogenic properties. It has been shown that various procedures diminish potency. These include treatment with ethylenediaminetetracetic acid (EDTA) and other chelators; high concentrations of hydrogen ions; fixative-decalcifying agents; gamma or ultraviolet radiation and high frequency ultrasonic radiation. The fundamental effects of such treatments upon the specimens have been studied, though no very definite conclusion seems to have been reached yet about the changes responsible for loss of osteogenic potency. Work with autoradiographs, entailing experiments with labelled implants, seems to suggest that whatever the nature of the osteogenic stimulus may be, it is not a diffusible macromolecular or specific chemical inducing agent.

Second, other experiments have been carried out to test the ability of various sites to respond to the osteogenic stimulus emitted by the decalcified lyophilized bone matrix. An attempt was made to express their response in a quantitative as well as a qualitative manner. Data are presented (Urist *et al.* 1969) about results with no less than 900 implants and it

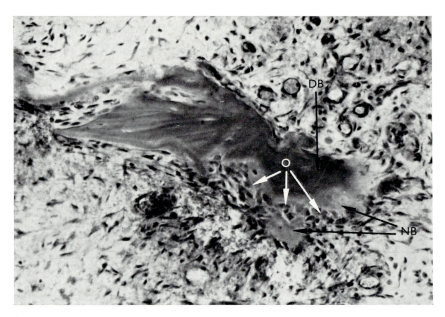

Fig. 79. The darker mass (DB) in the centre is a splinter of old dead bone encountered in a human fracture callus. Its bottom edge is covered by a layer of newly formed bone, which appears paler – staining (NB). Active osteoblasts (O) are easily identi-fied there. Vascular young connective tissue surrounds the bone. Magnification, × 256.

transpires that implant sites can be divided into three categories on the basis of their response. At one end of the scale are those which, it seems, do not ever produce bone; these include lymph nodes, thymus, adrenal, thyroid, parathyroid. At the other end, and giving a uniformly high success rate, are skeletal muscle, subcutaneous connective tissue, pericardium, lung, bone marrow, diaphyseal bone defect, kidney calyx. An intermediate group giving lower success rates and poorer yields includes, for example, bladder wall, brain, dermal tissue, mesentery, ovary, testis, pancreas. This work has been carried further more recently; it has been shown that preparations made from the histologically abnormal bone of animals suffering from lathyrism will not induce osteogenesis.

Millipore diffusion chambers have been used for other experiments with decalcified lyophilized bone matrix (Büring & Urist, 1967). It was found that, when minced muscle and the matrix were mixed and implanted within a chamber, cartilage or chondro-osteoid developed within and, on the outer surface of the membrane directly over the chondro-osteoid, new bone was produced. Chambers loaded with matrix or muscle alone did not lead to transfilter induction of bone. Recently, it has been claimed

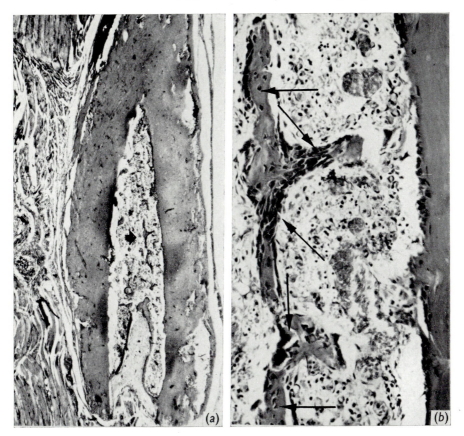

Fig. 80. (*a*) A piece of decalcified lyophilized long bone had been implanted intra-muscularly in an allogenic rat three weeks previously. It is seen cut longitudinally so that the original marrow cavity (C) is in the centre. Host muscle and connective tissue is seen on the left. The marrow cavity has been invaded by host cells. New bone has been induced to form in several places; the horizontal arrow marks one. Implant material kindly supplied by Dr M. Urist. Magnification, × 16.

(*b*) A higher magnification view of the area indicated by the arrow in (*a*). The lyophilized decalcified implanted bone matrix runs along the right hand side of the field; near the left is a trabeculum of newly-formed woven bone containing osteo-cytes (horizontal arrows); osteoblasts (oblique arrows) and an osteoclast (vertical arrow) are present. Magnification, × 160.

that cartilage can be induced even in tissue culture from minced muscle maintained in contact with pieces of the treated bone matrix (Urist & Nogami, 1970).

The present state of knowledge of this property of processed bone matrix can be summarized by the statement that bone inductions un-

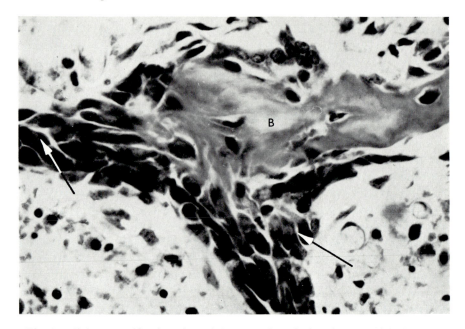

Fig. 81. Higher magnification view of the area of newly-forming bone (B) shown in Fig. 80 (*b*). Typical active osteoblasts can be seen with their darkly-staining, basophilic cytoplasm and clear juxtanuclear vacuoles (for example, arrows). Magnification, × 560.

doubtedly occur, but why they do so is unknown, although plenty of speculative explanations are available. The inability of certain sites to respond (or their poor response) also remains more or less enigmatical.

2. MATERIALS OF NON-BIOLOGICAL ORIGIN

Certain apparently practically inert, insoluble, non-biological materials turn out to emit an osteogenic stimulus. For example, following surgical removal of the breast, a prosthesis has sometimes been prepared by implanting a sponge of synthetic polymer. One of the materials which has been used for this purpose is polyhydroxyethylmethacrylate sponge. Experiments have shown that, in the pig at any rate, this material can induce the formation of bone (Winter, 1970; Winter & Simpson, 1969). On the other hand, polyurethane sponge did not induce bone in the guinea-pig even after many weeks (Beresford & Hancox, 1967). The explanation for this difference remains to be discovered.

3. CALCITONIN AND BONE FORMATION

Until about ten years ago it was generally thought that the regulation of the blood calcium level was controlled by a single hormone, namely, parathormone secreted by the parathyroid gland. It has been shown that the effect of the hormone was to mobilize calcium from bone to blood so that the level rose and it was assumed that the level returned to normal passively, as the output of parathyroid hormone fell.

Beginning in 1961, however, evidence began to accumulate (Copp *et al.*), for the existence of a second factor actively responsible for lowering the blood calcium. The subsequent history of the growth of knowledge about this factor, now known to be a polypeptide hormone and christened calcitonin, makes extremely interesting reading (see, for example, Rasmussen & Pechet, 1970; Hirsch & Munson, 1968; MacIntyre, 1967). Early controversy about which gland secretes calcitonin has been settled in favour of the thyroid, not the parathyroid as first suggested. The cells responsible for its secretion have been identified. Generally known as 'C' cells, they are epithelial cells adjacent to the ordinary thyroxine-secreting cells, but easily distinguishable from them by morphological, cytochemical and immunofluorescent methods. It has been shown that the C cells take embryological origin from the ultimobranchial body. In mammals, the latter eventually loses its C cells to the thyroid, in which they disperse, but in birds and fishes the body retains its morphological identity (and C cells) and is a potent source for the extraction of calcitonin. Highly purified preparations are now available; the amino acid sequence of the chain has been established and the hormone has been synthesized.

Thyrocalcitonin seems to produce its specific effect, at any rate in part, by a direct action on bone. How it does this is uncertain but one or other or both of the two following possibilities must be involved: either a shift of calcium from bone to blood is prevented through inhibition of resorption (for which there is a good deal of experimental evidence), or calcium is shifted from blood to bone. In the latter instance, the additional calcium load must be borne either by bone tissue already in existence, or by the rapid laying down of new bone. Here one recalls the medullary bone laid down as a calcium store by birds in the egg-laying cycle (see p. 31). If it were to be shown that calcitonin produces new bone, then the hormone could be added to the list of substances apparently capable of producing an osteogenic stimulus. There are, in fact, three lines of evidence which suggest that calcitonin indeed has such an effect. First, administration of calcitonin to intact animals has been claimed to lead to increase in the thickness of bones (Wase *et al.* 1967). Second, experiments with bone in tissue culture (Gaillard, 1970) suggest that, in the presence of calcitonin, osteoblastic activity and osteogenesis are increased. Third, Thompson & Urist (1970) claim that more ectopic bone forms around demineralized bone implants in animals injected with calcitonin than in controls. It will be interesting to see if this effect of calcitonin upon osteoblasts can be substantiated by further experiments.

REFERENCES

ABELSON, P. H. (1956). Paleobiochemistry. *Scientific American*, **195**, 2, 83–92.

ACKERMAN, N. D. (1968). Bone formation in the gastric mucosa following subtotal gastrectomy in rats. *Expl Molec. Path.* **9**, 125–30.

AEGETER, E. & KIRKPATRICK, J. A. (1963). *Orthopedic Diseases*, 2nd edn. Philadelphia: W. B. Saunders.

ALGIRE, G. H. (1957). Diffusion chamber techniques for studies of cellular immunity. *Ann. N.Y. Acad. Sci.* **69**, 663–9.

AMPRINO, R. (1963). On the growth of cortical bone and the mechanism of osteon formation. *Acta anat.* **52**, 177–87.

AMPRINO, R. & MAROTTI, G. (1964). A topographic quantitative study of bone formation and reconstruction. In *Bone and Tooth* (ed. H. J. J. Blackwood), pp. 21–33. London: Pergamon Press.

ANDERSON, C. E. & PARKER, J. (1968). Electron microscopy of the epiphyseal cartilage plate. *Clin. Orthop.* **58**, 225–41.

ANDERSON, D. J., HANNAM, A. G. & MATTHEWS, B. (1970). Sensory mechanisms in mammalian teeth and their supporting structures. *Physiol. Rev.* **50**, 171–95.

ANDERSON, H. C. (1967). Electron microscope studies of induced cartilage development and calcification. *J. Cell Biol.* **35**, 81–101.

ANDERSON, H. C., MERKER, P. C. & FOGH, J. (1964). Formation of tumours containing bone after intramuscular injection of transformed human amnion cells (FL) into cortisone-treated mice. *Am. J. Path.* **44**, 507–19.

ARMSTRONG, W. G. & TARLO, L. B. H. (1966). Amino-acid components in fossil calcified tissues. *Nature, Lond.* **210**, 481–2.

ARNOLD, J. S. & GEE, W. S. S. (1957). Bone growth and osteoclastic activity as indicated by radioautographic distribution of plutonium. *Am. J. Anat.* **101**, 367–417.

ASCENZI, A., BONUCCI, E. & BOCCIARELLI, S. (1967). An electron microscope study on primary periosteal bone. *J. Ultrastruct. Res.* **18**, 605–18.

BALAZS, A. (1965). Amino sugar containing macromolecules in the tissues of the eye and ear. In *The Amino Sugars 2A* (eds. E. A. Balzacs and R. W. Jeanloz), pp. 141–60. London: Academic Press.

BARNICOT, N. A. (1941). Studies on the factors involved in bone absorption. *Am. J. Anat.* **68**, 497–531.

BARNICOT, N. A. (1945). Some data on the effect of parathormone on the grey-lethal mouse. *J. Anat.* **79**, 83–91.

BARNICOT, N. A. (1947). The supravital staining of osteoclasts with neutral red: their distribution in the parietal bone of normal growing mice, and a comparison with the mutants grey-lethal and hydrocephalus 3. *Proc. Roy. Soc.* B **134**, 467–85.

BARNICOT, N. A. (1948). The local action of the parathyroid and other tissues on bone in intracerebral grafts. *J. Anat.* **82**, 233–68.

BASSETT, C. A. L. (1968). Biologic significance of piezoelectricity. *Calc. Tiss. Res.* **1**, 252–72.

BASSETT, C. A. L., PAWLUK, R. J. & BECKER, R. O. (1964). Effects of electric currents on bone *in vivo*. *Nature, Lond.* **204**, 652–4.

BAUD, C. A. (1966). The fine structure of normal and parathormone-treated bone cells. *Fourth European Symposium on Calcified Tissues*, abridged proceedings (eds. P. J. Gaillard, A. van den Hoof and R. Steendijk), pp. 4–6. Amsterdam: Excerpta Medica Foundation.

BAUD, C. A. (1968). Submicroscopic structure and functional aspects of the osteocyte. *Clin. Orthop.* **56**, 227–36.

BELANGER, L. F. (1969). Osteocytic osteolysis. *Calc. Tiss. Res.* **4**, 1–12.

BELANGER, L. F. & MIGICOVSKY, D. H. (1963). Histochemical evidence of proteolysis in bone: the influence of parathormone. *J. Histochem. Cytochem.* **11**, 734–7.

BELANGER, L. F., BELANGER, C. & SEMBA, T. (1967). Technical approaches leading to the concept of osteocytic osteolysis. *Clin. Orthop.* **54**, 187–96.

BELANGER, L. F., ROBICHON, J., MIGIKOVSKY, B. B., COPP, D. H. & VINCENT, J. (1963). In *Mechanisms of Hard Tissue Destruction* (ed. R. F. Sognnaes), pp. 551–6. Washington: Am. Assoc. Adv. Sci. publ. no. 75.

BELANGER, L. F., SEMBA, T., TOLNAI, S., COPP, D. H., KROOL, L. & GREIS, C. (1966). The two faces of resorption. In *Calcified Tissues* (eds. H. Fleisch, H. J. J. Blackwood and Maureen Owen), pp. 1–10. Berlin: Springer.

BENTLEY, F. H. (1936). Wound healing *in vitro*. *J. Anat.* **70**, 499–538.

BERESFORD, W. A. & HANCOX, N. M. (1967). Urinary bladder mucosa and bone regeneration in guinea pig and rat. *Acta anat.* **66**, 78–117.

BERNARD, G. W. & PEASE, D. C. (1969). An electron microscope study of initial intramembranous osteogenesis. *Am. J. Anat.* **125**, 271–90.

BERNFIELD, M. R. (1970). Collagen synthesis during epitheliomesenchymal interactions. *Devl Biol.* **22**, 213–31.

BERRILL, N. J. (1955). *The Origin of Vertebrates*. London: Oxford University Press.

BINGHAM, P. J., BRAZELL, I. A. & OWEN, M. (1969). The effect of parathyroid extract on cellular activity and plasma calcium levels *in vivo*. *J. Endocr.* **45**, 387–400.

BINGHAM, P. J. & OWEN, M. (1968). Effects of PTE on bone metabolism *in vivo*. *Calc. Tiss. Res.* **2** (suppl.), 46.

BLOOM, W. & BLOOM, MARGARET. (1940). Calcification and ossification. Calcification of developing bones in embryonic and newborn rats. *Anat. Rec.* **78**, 497–541.

BLOOM, W., BLOOM, M. A. & McLEAN, F. C. (1941). Calcification and ossification. Medullary bone changes in the reproductive cycle of female pigeons. *Anat. Rec.* **87**, 443–75.

BOCCIARELLI, D. S. (1970). Morphology of crystallites in bone. *Calc. Tiss. Res.* **5**, 261–9.

BOOTHROYD, B. (1964). The problem of demineralisation in thin sections of fully calcified bone. *J. Cell Biol.* **20**, 165–73.

BORNSTEIN, P. & PIEZ, K. A. (1964). A biochemical study of human skin collagen and the relation between intra- and inter-molecular cross-linking. *J. clin. Invest.* **43**, 1813–23.

BOROJEVIC, R., FRY, W. A., JONES, W. C., LEVI, C., RASMONT, R., SARA, M. & VACELET, J. (1967). Mise au point actuelle de la terminologie des éponges. *Bull. Mus. Hist. nat., Paris,* **39**, 1224–35.

BOWNESS, J. M. (1968). Present concepts of the role of ground substance in calcification. *Clin. Orthop.* **59**, 233–47.

BOYDE, A. & HOBDELL, M. (1968). Scanning electron microscopy of bone. *Calc. Tiss. Res.* **2** (suppl.), 4–4A.

BOYDE, A. & HOBDELL, M. (1969). Scanning electron microscopy of lamellar bone. *Z. Zellforsch. mikrosk. Anat.* **93**, 213–31.

BRADLEY, S. (1970). An analysis of self-differentiation of chick limb buds in chorioallantoic grafts. *J. Anat.* **107**, 479–90.

British Medical Journal. (1970). Prolonged weightlessness. Editorial, **3**, 4.

Brown, W. E. (1966). Crystal growth of bone mineral. *Clin. Orthop.* **44**, 205–20.

Büring, K. & Urist, M. R. (1967). Transfilter bone induction. *Clin. Orthop.* **54**, 235–242.

Burkhart, J. M. & Jowsey, J. (1967). Parathyroid and thyroid hormones in the development of immobilization osteoporosis. *Endocrinology*, **81**, 1053–62.

Cameron, D. A. (1963). The fine structure of bone and calcified cartilage. *Clin. Orthop.* **26**, 199–228.

Cameron, D. A. (1968). The Golgi apparatus in bone and cartilage cells. *Clin. Orthop.* **58**, 191–211.

Cameron, D. A. & Robinson, R. A. (1958). The presence of crystals in the cytoplasm of large cells adjacent to sites of bone absorption. *J. Bone Jt Surg.* **40A**, 414–18.

Cameron, D. A., Paschall, H. A. & Robinson, R. A. (1964). The ultrastructure of bone cells. In *Bone Biodynamics* (ed. H. M. Frost), pp. 91–104. Boston: Little, Brown & Co.

Cameron, D. A., Paschall, H. A. & Robinson, R. A. (1967). Changes in the fine structure of bone cells after the administration of parathyroid extract. *J. Cell Biol.* **33**, 1–13.

Campo, R. D. (1970). Protein–polysaccharides of cartilage and bone in health and disease. *Clin. Orthop.* **68**, 182–209.

Campo, R. D. & Dziewiatkowski, D. D. (1963). Turnover of the organic matrix of cartilage and bone as visualised by autoradiography. *J. Cell Biol.* **18**, 19–29.

Campo, R. D. & Tourtellotte, C. D. (1967). The composition of bovine cartilage and bone. *Biochim. biophys. Acta* **141**, 614–24.

Carneiro, J. & Leblond, C. P. (1959). Role of osteoblasts and odontoblasts in secreting the collagen of bone and dentin as shown by radioautography in mice given tritium-labelled glycine. *Expl Cell Res.* **18**, 291–300.

Chang, H. Y. (1951). Grafts of parathyroid and other tissues to bone. *Anat. Rec.* **111**, 23–70.

Chapman, G. (1966). The structure and functions of the mesoglea. In *The Cnidaria and their Evolution.* (ed. W. J. Rees), pp. 147–68. New York: Academic Press.

Cohen, J. & Harris, W. H. (1958). The three-dimensional anatomy of Haversian systems. *J. Bone Jt Surg.* **40A**, 419–34.

Collins, D. (1966). *Pathology of Bone.* London: Butterworth.

Cooper, R. R., Milgram, J. W. & Robinson, R. A. (1966). Morphology of the osteon. *J. Bone Jt Surg.* **48A**, 1239–71.

Cooper, D. R. & Russell, A. E. (1969). Intra- and intermolecular crosslinks in collagen in tendon, cartilage and bone. *Clin. Orthop.* **67**, 188–209.

Copp, D. H., Cameron, E. C., Cheney, B. A., Davidson, A. G. F. & Henze, K. G. (1962). Evidence for calcitonin – a new hormone from the parathyroid that lowers blood calcium. *Endocrinology* **70**, 638–49.

Copp, D. H., Davidson, A. G. F. & Cheney, B. A. (1961). Evidence for a new parathyroid hormone which lowers blood calcium. *Proc. Can. fed. biol. Soc.* **4**, 17–20.

Copp, D. H., Low, B. S., O'Dor, R. K. & Parkes, C. O. (1969). Calcitonin in non-mammals. *Calc. Tiss. Res.* **2** (suppl.), 29.

Cornah, M. S., Meachim, G. & Parry, E. W. (1970). Banded structures in the matrix of human and rabbit nucleus pulposus. *J. Anat.* **107**, 351–62.

Cox, R. W., Grant, R. A. & Horne, R. W. (1967). The structure and assembly of collagen fibrils. *J. Roy. microsc. Soc.* **87**, 123–42.

Crang, R. E., Holsen, R. C. & Hitt, J. D. (1968). Calcite production in mitochondria of earthworm calciferous glands. *Am. Inst. biol. Sci. Bull.* **18**, 299–308.

180   References

CRAWFORD, G. N. C. (1940). The evolution of the Haversian pattern in bone. *J. Anat.* **74**, 284–99.

DANIELLI, J. F. (1942). The cell surface and cell physiology. In *Cytology and Cell Physiology* 1st edn (ed. G. H. Bourne), pp. 68–93. Oxford: Clarendon Press.

DANIS, A. (1957). Étude de l'ossification dans les greffes de moelle osseuse. *Acta chir. belg.* (suppl.), **3**, 1–120.

DECKER, J. D. (1966). An electron microscope investigation of osteogenesis in the embryonic chick. *Am. J. Anat.* **118**, 591–613.

DENISON, R. H. (1963). The early history of the vertebrate calcified skeleton. *Clin. Orthop.* **31**, 141–51.

DINGLE, J. T. & FELL, H. B. (1969). *Lysosomes in Biology and Pathology.* London: North-Holland Publishing Co.

DOBELL, C. (1932). *Antony van Leeuwenhoek and his Little Animals.* London: John Bale, Sons & Danielsson, Ltd.

DOBERENZ, A. R. (1967). Ultrastructure of fossil dentinal collagen. *Calc. Tiss. Res.* **1**, 166–9.

DORRIS, F. (1935). The development of structure and function in the digestive tract of *Amblystoma punctatum. J. exp. Zool.* **70**, 491–527.

DUDLEY, R. H. & SPIRO, D. (1961). The fine structure of bone cells. *J. biophys. biochem. Cytol.* **11**, 627–71.

EANES, E. D., HARPER, R. A., GILLESSEN, I. H. & POSNER, A. S. (1966). An amorphous component in bone mineral. *Fourth European Symposium on Calcified Tissues*, abridged proceedings (eds. P. J. Gaillard, A. van den Hoof and R. Steendijk), pp. 24–6. Amsterdam: Excerpta Medica Foundation.

EANES, E. D., GILLESSEN, I. H. & POSNER, A. S. (1965). Intermediate states in the precipitation of hydroxyapatite. *Nature, Lond.* **208**, 365–7.

EANES, E. D. & POSNER, A. S. (1965). Kinetics and mechanism of conversion of non-crystalline calcium phosphate to crystalline hydroxyapatite. *Trans. N.Y. Acad. Sci.* **28**, 233–41.

EANES, E. D. & POSNER, A. S. (1970). Structure and chemistry of bone collagen. In *Biological Calcification. Cellular and Molecular Aspects* (ed. H. Schraer), pp. 1–26. Amsterdam: North-Holland Publishing Co.

EASTOE, J. E. (1968). Chemical aspects of the matrix concept in calcified tissue organisation. *Calc. Tiss. Res.* **2**, 1–19.

EASTOE, J. E. & EASTOE, B. (1954). The organic constituents of mammalian compact bone. *Biochem. J.* **57**, 453–9.

EBNER, V. V. (1875). Über den feineren Bar der Knochensubstanz. *Sber. Akad. Wiss. Wien, Math. Naturw. Kl. Abt.* 72, 49.

EBNER, V. V. (1894). Über eine optische Reaktion der Bindesubstanzen auf Phenole. *Sber. Akad. Wiss. Wien* **103**, *Abt.* 3, 162.

EDEIKEN, J. & HODER, P. J. (1967). *Roentgen Diagnosis of Diseases of Bone.* Baltimore: Williams & Wilkins Co.

ENGSTRÖM, A. (1960). The structure of bone: an excursion into molecular biology. *Clin. Orthop.* **17**, 34–7.

ENLOW, D. H. (1963). *Principles of Bone Remodelling.* Springfield: Charles Thomas.

ENLOW, D. H. & BROWN, S. O. (1956). A comparative histological study of fossil and recent bone tissues. Part 1. *Texas J. Sci.* **8**, 405–43.

ENLOW, D. H. & BROWN, S. O. (1957). A comparative histological study of fossil and recent bone tissues. Part 2. *Texas J. Sci.* **9**, 186–214.

ENLOW, D. H. & BROWN, S. O. (1958). A comparative histological study of fossil and recent bone tissues. Part 3. *Texas J. Sci.* **10**, 187–230.

EVANS, F. G. & BANG, S. (1966). Physical and histological differences between human fibular and femoral compact bone. In *Studies on the Anatomy and Function of Bone and Joints* (ed. F. Gaynor Evans), pp. 142–50. New York: Springer.

FACCINI, J. M. (1969). Fluoride and bone. *Calc. Tiss. Res.* **3**, 1–16.

FELL, H. B. (1932). The osteogenic capacity *in vitro* of periosteum and endosteum isolated from the limb skeleton of fowl embryos and young chicks. *J. Anat.* **66**, 157–80.

FELL, H. B. (1933). Chondrogenesis in cultures of endosteum. *Proc. Roy. Soc.* B **112**, 417–27.

FELL, H. B. & DINGLE, J. T. (1969). Endocytosis of sugars in embryonic skeletal tissues in organ culture. 1. General introduction and histological effects. *J. Cell Sci.* **4**, 89–103.

FERNANDEZ-MADRID, F. (1970). Collagen biosynthesis. A review. *Clin. Orthop.* **68**, 103–81.

FIORE, J. M. & SMYTH, W. T. (1960). Massive osteolysis of bone: report of a fatal case with temporary reconstitution of the affected bone following irradiation. *Ann. intern. Med.* **53**, 807–16.

FISCHMAN, D. A. & HAY, E. D. (1962). Origin of osteoclasts from mononuclear leukocytes in regenerating newt limbs. *Anat. Rec.* **143**, 329–34.

FITTON JACKSON, S. (1957). The fine structure of developing bone in the embryonic fowl. *Proc. Roy. Soc.* B **146**, 270–80.

FITTON JACKSON, S. & SMITH, R. H. (1957). Studies on the biosynthesis of collagen. 1. The growth of fowl osteoblasts and the formation of collagen in tissue culture. *J. biophys. biochem. Cytol.* **3**, 897–929.

FLEISCH, H. (1964). Role of nucleation and inhibition in calcification. *Clin. Orthop.* **32**, 170–80.

FRAME, B. & NIXON, R. K. (1968). Bone marrow mast cells in osteoporosis of aging. *New Engl. J. Med.* **279**, 626–31.

FRANK, R. M. (1966). Ultrastructure of human dentine. Calcified Tissues 1965. *Proceedings 3rd European Symposium on Calcified Tissues.* (eds. H. Fleisch, H. J. J. Blackwood and M. Owen), pp. 259–72. New York: Springer.

FRIEDENSTEIN, A. Y. (1962). Humoral nature of osteogenic activity of transitional epithelium. *Nature, Lond.* **194**, 698–9.

FRIEDENSTEIN, A. Y. (1968). Induction of bone tissue by transitional epithelium. *Clin. Orthop.* **59**, 21–37.

FRIEDENSTEIN, A. Y., LALYKINA, K. S. & TOLMACHEVA, A. A. (1967). Osteogenic activity of periosteal fluid cells induced by transitional epithelium. *Acta anat.* **68**, 532–49.

FRIEDMAN, B., HEIPLE, K. G., VESSELY, J. C. & HANAOKA, H. (1968). Ultrastructural investigation of bone induction by an osteosarcoma, using diffusion chambers. *Clin. Orthop.* **59**, 39–57.

FROST, H. M. (1963). *Bone Remodelling Dynamics.* Springfield: Charles Thomas.

FROST, H. M. (1964). *Bone Biodynamics.* Boston: Little, Brown & Co.

FROST, H. M. (1969). Tetracycline-based histological analysis of bone remodelling. *Calc. Tiss. Res.* **3**, 211–37.

GAILLARD, P. J. (1955). Parathyroid gland tissue and bone *in vitro*. II. *Proc. K. ned. Akad. Wet.* C **58**, 279–83.

GAILLARD, P. J. (1959). Parathyroid gland and bone *in vitro*. *Devl Biol.* **1**, 1, 152–71.

GAILLARD, P. J. (1961). Parathyroid and bone in tissue culture. In *The Parathyroids* (eds. R. O. Greep and R. V. Talmage), p. 20. Springfield: Charles Thomas.

GAILLARD, P. J. (1970). Induction of bone formation in explanted bone rudiments by calcitonin and imidazole. *Calc. Tiss. Res.* **4** (suppl.), 86–7.

182 *References*

GEBHARDT, F. (1906). Ueber functionell wichtigen Anordnungsweise der grösere und feineren Banelemente des Wirbeltierknochens. *Arch. EntwMech. Org.* **20**, 187–334.

GLASSTONE, S. (1936). The development of tooth germs *in vitro*. *J. Anat.* **70**, 260–301.

GLAUERT, A., FELL, H. B. & DINGLE, J. T. (1969). Endocytosis of sugars in embryonic skeletal tissues in organ culture. II. Effect of sucrose on cellular fine structure. *J. Cell Sci.* **4**, 105–31.

GLIMCHER, M. J. (1959). Molecular biology of mineralised tissues with particular reference to bone. *Rev. mod. Phys.* **31**, 359–93.

GLIMCHER, M. J., FRANCOIS, C. J., RICHARDS, L. & KRANE, S. M. (1964). The presence of organic phosphorus in collagens and gelatins. *Biochim. biophys. Acta* **93**, 585–602.

GLIMCHER, M. J., HODGE, A. J. & SCHMITT, F. O. (1957). Macromolecular aggregation states in relation to mineralisation: the collagen–hydroxyapatite system as studied *in vitro*. *Proc. natn. Acad. Sci. U.S.A.* **43**, 860–7.

GLIMCHER, M. J. & KATZ, E. P. (1965). The organisation of collagen in bone: the role of non-covalent bonds in the relative insolubility of bone collagen. *J. Ultrastruct. Res.* **12**, 705–29.

GLIMCHER, M. J. & KRANE, S. M. (1968). Organisation and structure of bone. In *Treatise on Collagen, 2B* (ed. B. S. Gould), pp. 68–251. London and New York: Academic Press.

GLÜCKSMANN, A., HOWARD, A. & PELC, S. R. (1956). The uptake of radioactive sulfate by cells, fibres and ground substance of mature and developing connective tissue in the adult mouse. *J. Anat.* **90**, 478–85.

GOLDHABER, P. (1960). Behavior of bone in tissue culture. In *Calcification in Biological Systems* (ed. R. F. Sognnaes), pp. 349–72. Washington: Am. Assoc. Adv. Sci. publ. no. 64.

GOLDHABER, P. (1961). Osteogenic induction across millipore filters *in vivo*. *Science, N.Y.* **133**, 2065–7.

GOLDHABER, P. (1965). Bone resorption factors, cofactors and giant vacuole osteoclasts in tissue culture. In *The Parathyroid Glands* (eds. P. J. Gaillard, R. V. Talmage and A. M. Budy), pp. 153–71. Chicago: University of Chicago Press.

GONZALES, F. (1961). Electron microscopy of osteoclasts. *Anat. Rec.* **139**, 330–49.

GONZALES, F. & KARNOVSKY, M. J. (1961). Electron microscopy of osteoclasts in healing fractures of rat bone. *J. biophys. biochem. Cytol.* **9**, 299–316.

GOODSIR, J. & GOODSIR, H. D. S. (1845). *Anatomical and Pathological Observations.* Edinburgh: Macphail.

GRANT, R. A., HORNE, R. W. & COX, R. W. (1965). New model for the tropocollagen molecule and its mode of aggregation. *Nature, Lond.* **207**, 822–4.

GREENAWALT, J. W., ROSSI, C. S. & LEHINGER, A. L. (1964). Effect of active accumulation of calcium and phosphate ions on the structure of rat liver mitochondria. *J. Cell Biol.* **23**, 21–35.

GRIFFITH, G. C., NICHOLS, G., ASHER, J. D. & FLANAGAN, B. (1965). Heparin osteoporosis. *J. Am. med. Ass.* **193**, 91–4.

GROBSTEIN, C. (1968). Developmental significance of interface materials in epithelio-mesenchymal interaction. In *Epithelial-mesenchymal Interactions* (eds. R. Fleischmajer and R. E. Billingham), pp. 173–6. Baltimore: Williams & Wilkins Co.

GROSS, J., HIGHBERGER, J. H. & SCHMITT, F. O. (1954). Extraction of collagen from connective tissue by neutral salt solutions. *Proc. natn. Acad. Sci. U.S.A.* **41**, 1–7.

GROSS, J. & PIEZ, K. A. (1960). The nature of collagen. 1. Invertebrate collagens. In *Calcification in Biological Systems* (ed. R. F. Sognnaes), pp. 395–411. Washington: Am. Assoc. Adv. Sci. publ. no. 64.

GROSSFELD, H., MEYER, K. & GODMAN, G. C. (1956). Acid mucopolysaccharides produced in tissue culture. *Anat. Rec.* **124**, 489–515.

GRÜNEBERG, H. (1936). Grey-lethal, a new mutation in the house mouse. *J. Hered.* **27**, 105–9.

HALLIDAY, D. R., DAHLIN, D. C., PUGH, D. G. & YOUNG, H. (1964). Massive osteolysis and angiomatosis. *Radiology*, **82**, 637–44.

HALSTEAD, L. B. (1969*a*). Are mitochondria directly involved in biological mineralisation? *Calc. Tiss. Res.* **3**, 103–4.

HALSTEAD, L. B. (1969*b*). Calcified tissues in the earlier vertebrates. *Calc. Tiss. Res.* **3**, 107–24.

HALSTEAD, L. B. (1969*c*). *The Pattern of Vertebrate Evolution*. Edinburgh: Oliver & Boyd.

HANCOX, N. M. (1949). The osteoclast. *Biol. Rev.* **24**, 448–67.

HANCOX, N. M. (1956). The osteoclast. In *The Biochemistry and Physiology of Bone*. (ed. G. H. Bourne), pp. 213–47. London: Academic Press.

HANCOX, N. M. (1963*a*). Osteoclasts. In *The Biology of Cells and Tissues in Culture* (ed. E. N. Willmer), pp. 262–72. London: Academic Press.

HANCOX, N. M. (1963*b*). Motion picture studies of osteoclasts. In *Cinemicrography in Cell Biology* (ed. G. G. Rose), pp. 141–90. New York: Academic Press.

HANCOX, N. M. & BOOTHROYD, B. (1961). Motion picture and electron microscope studies on the embryonic avian osteoclast. *J. biophys. biochem. Cytol.* **11**, 651–61.

HANCOX, N. M. & BOOTHROYD, B. (1963). Structure–function relationships in the osteoclast. In *Mechanisms of Hard Tissue Destruction* (ed. R. F. Sognnaes), pp. 497–514. Washington: Am. Assoc. Adv. Sci. publ. no. 75.

HANCOX, N. M. & BOOTHROYD, B. (1964). Ultrastructure of bone formation and resorption. In *Modern Trends in Orthopaedics. 4. Science of Fractures* (ed. J. M. P. Clarke), pp. 26–52. London: Butterworth.

HANCOX, N. M. & BOOTHROYD, B. (1965). Electron microscopy of the early stages of osteogenesis. *Clin. Orthop.* **40**, 153–61.

HATTNER, R. S. & McMILLAN, D. E. (1968). Influence of weightlessness upon the skeleton: a review. *Aerospace Medicine*, **39**, 849–60.

HAVERS, C. (1692). Some new observations on bone. Facsimile reproduction. In *Principles of Bone Remodelling* by D. H. Enlow. Springfield: Charles Thomas, 1963.

HEANEY, R. P. (1964). Disuse osteoporosis. In *Dynamic Studies of Metabolic Bone Disease* (eds. O. H. Pearson and G. F. Joplin), pp. 77–85. Oxford: Blackwell Scientific Publications.

HEINEN, J. H. (1952). Fig. 119. In Maximow and Bloom, *Textbook of Histology*, 5th edn. Philadelphia: W. B. Saunders, 1948.

HELLER-STEINBERG, M. (1951). Ground substance, bone salts, and cellular activity in bone formation and destruction. *Am. J. Anat.* **89**, 347–72.

HELLER, M., McLEAN, F. C. & BLOOM, W. (1950). Cellular transformations in mammalian bones induced by parathyroid extract. *Am. J. Anat.* **87**, 315–48.

HERMAN, H. & DALLEMAGNE, M. J. (1961). The main mineral constituent of bone and teeth. *Archs oral Biol.* **5**, 137–44.

HERRING, G. M. (1964). Chemistry of the bone matrix. *Clin. Orthop.* **36**, 169–83.

HILFER, S. R. (1968). Cellular interactions in the genesis and maintenance of thyroid characteristics. In *Epithelial-mesenchymal Interactions* (eds. R. Fleischmajer and R. E. Billingham), pp. 177–99. Baltimore: Williams & Wilkins Co.

HIRSCH, P. F. & MUNSON, P. L. (1968). Thyrocalcitonin. *Physiol. Rev.* **49**, 548–622.

HO, T. Y. (1965). The amino acid composition of bone and tooth proteins in late Pleistocene mammals. *Proc. natn. Acad. Sci. U.S.A.* **54**, 26–31.

Ho, T. Y. (1967). The amino acids of bone and dentine collagens in Pleistocene mammals. *Biochim. biophys. Acta* **133**, 568–73.

Hodge, A. J. & Petruska, J. A. (1963). Recent studies with the electron microscope on ordered aggregates of the tropocollagen macromolecule. In *Aspects of Protein Structure* (ed. G. N. Ramachandran), pp. 289–300. London and New York: Academic Press.

Hodge, A. J., Petruska, J. A. & Bailey, A. J. (1965). The subunit structure of the tropocollagen macromolecule and its relation to various ordered aggregation states. In *Structure and Function of Connective and Skeletal Tissue* (eds. S. Fitton Jackson, R. D. Harkness, S. M. Partridge and G. R. Tristram), pp. 31–41. London: Butterworth.

Höhling, H. J. (1969). Collagen mineralisation in bone, dentine, cementum and cartilage. *Naturwissenschaften*, **56**, 466.

Hong, Y. C., Yen, P. K. & Shaw, J. H. (1968a). An analysis of the growth of the cranial vault in rabbits by vital staining with lead acetate. *Calc. Tiss. Res.* **2**, 271–85.

Hong, Y. C., Yen, P. K. & Shaw, J. H. (1968b). Microscopic evaluation of the effects of some vital staining agents on growing bones in rabbits. *Calc. Tiss. Res.* **2**, 286–95.

Horne, R. W. (1965). Negative staining methods. In *Techniques for Electron Microscopy*, 2nd edn (ed. D. Kay), pp. 328–55. Oxford: Blackwell Scientific Publications.

Howship, J. (1817). Observations on the morbid structure of bone. *Med. Chir. Trans.* VIII, 57–107.

Huggins, C. B. (1931). The formation of bone under the influence of epithelium of the urinary tract. *Archs Surg., Chicago*, **22**, 377–408.

Hunter, D. & Turnbull, H. N. (1931). Hyperparathyroidism: generalised osteitis fibrosa. *Br. J. Surg.* **19**, 203–84.

Hyslop, D. B. (1952). The effect of supravital dyes on osteoclasts in tissue culture. M.Sc. thesis, University of Liverpool.

Irving, J. T. & Wuthier, R. E. (1968). Histochemistry and biochemistry of calcification with special reference to the role of lipids. *Clin. Orthop.* **56**, 237–60.

Isaacs, W. A., Little, K., Currey, J. D. & Tarlo, L. B. (1963). Collagen and a cellulose-like substance in fossil dentine and bone. *Nature, Lond.* **197**, 192.

Jackson, D. S. & Bentley, J. P. (1960). On the significance of the extractable collagens. *J. biophys. biochem. Cytol.* **7**, 37–42.

Jande, S. S. & Belanger, L. F. (1969). Ultrastructural changes associated with osteocytic osteolysis in normal trabecular bone. *Anat. Rec.* **163**, 204–34.

Jaques, P. (1965). Studies on bone enzymes. The activation and release of latent acid hydrolases and catalase in bone-tissue homogenates. *Biochem. J.* **97**, 393–402.

Jee, W. S. S. & Nolan, P. D. (1963). Origin of osteoclasts from the fusion of phagocytes. *Nature, Lond.* **200**, 225–6.

Johnson, L. C. (1960). Mineralization of turkey leg tendon. 1. Histology and histochemistry of mineralization. In *Calcification in Biological Systems* (ed. R. F. Sognnaes), pp. 117–28. Washington: Am. Ass. Adv. Sci. publ. no. 64.

Jones, A. R. (1969). Mitochondria, calcification, and waste disposal. *Calc. Tiss. Res.* **3**, 363–5.

Jones, W. C. (1967). Sheath and axial filament of calcareous sponges. *Nature, Lond.* **214**, 365–8.

Jones, W. C. (1970). The composition, development, form and orientation of calcareous sponge spicules. *Symp. zool. Soc. Lond.* **25**, 91–123.

Jones, W. C. & James, D. W. F. (1969). An investigation of some calcareous sponge spicules by means of electron probe micro-analysis. *Micron*, **1**, 34–9.

JOWSEY, J. (1955). The use of the milling machine for preparing bone sections for microradiography and microautoradiography. *J. scient. Instrum.* **32**, 159.

JOWSEY, J. (1960). Age changes in human bone. *Clin. Orthop.* **17**, 210–17.

JOWSEY, J., KELLY, P. J., RIGGS, L., BIANCO, A. J., SCHOLZ, D. A. & GERSHON-COHEN, J. (1965). Quantitative microradiographic study of normal and osteoporotic bone. *J. Bone Jt Surg.* A **47**, 785–806.

JUVA, K. D., PROCKOP, J., COOPER, G. W. & LASH, J. W. (1966). Hydroxylation of proline and the intracellular accumulation of a polypeptide precursor of collagen. *Science, N.Y.* **152**, 92.

KATZ, E. P. (1969). The kinetics of mineralisation *in vitro*. I. The nucleation properties of 640 Å collagen at 25 °C. *Biochim. biophys. Acta*, **194**, 121–9.

KAY, M. I., YOUNG, R. A. & POSNER, A. S. (1964). Crystal structure of hydroxyapatite. *Nature, Lond.* **204**, 1050–2.

KIRBY-SMITH, H. T. (1933). Bone growth studies: a miniature bone fracture observed microscopically in a transparent chamber introduced into the rabbit's ear. *Am. J. Anat.* **53**, 377–402.

KLEBS, A. (1869). Die Einschmelzungs-Methode. *Arch. mikrosk. Anat. EntwMech.* **5**, 164–6.

KLEIN, E. (1883). *Elements of Histology*. London: Cassell & Co.

KLEIN, E. & NOBLE SMITH, E. (1880). *Atlas of Histology*. London: Smith, Elder & Co.

KNESE, K.-H. (1964). A histochemical study of the polysaccharides in osteogenic areas. In *Bone and Tooth* (ed. H. J. J. Blackwood), pp. 283–7. London: Pergamon Press.

KOJIMA, M. & OGATA, M. (1960). On the nature of the so-called osteoclasts. *Tohoku J. exp. Med.* **71**, 373–84.

KÖLLIKER, A. (1859). On the different types in the microstructure of the skeleton of osseous fishes. *Proc. Roy. Soc. Lond.* **9**, 656–68.

KÖLLIKER, A. (1873). *Die normale Resorption des Knochengewebes und ihre Bedeutung für die Entstehung der typischen Knochenformen*. Leipzig: Vogel.

LANE, N. J. (1968). Lipochondria neutral red granules and lysosomes: synonymous terms? In *Cell Structure and its Interpretation* (eds. S. M. McGee-Russell and K. F. A. Ross), pp. 169–82. London: Edward Arnold.

LASH, J. W. (1968). Somitic mesenchyme and its response to cartilage induction. In *Epithelial-mesenchymal Interactions* (eds. R. Fleischmajer and R. E. Billingham), pp. 165–72. Baltimore: Williams & Wilkins Co.

LAVINE, L. S., LUSTRIN, I. & SHAMOS, M. H. (1968). The influence of direct current *in vitro*. *Calc. Tiss. Res.* **2** (suppl.), 9.

LEEUWENHOEK, VAN A. (1693). An extract of a letter from Mr Anthony van Leeuwenhoek, containing several observations on the texture of the bones of animals compared with that of wood: on the bark of trees: on the little scales formed on the cuticula, & etc. *Phil. Trans. Roy. Soc.* **17**, 838.

LEONARD, F. & SCULLIN, R. I. (1969). New mechanism for calcification of skeletal tissues. *Nature, Lond.* **224**, 1113–15.

LERICHE, A. & POLICARD, A. (1928). *The Normal and Pathological Physiology of Bone*. Eng. trans. by S. Moore and A. S. Kay. St Louis: Mosby.

LEVANDER, G. (1964). *Induction Phenomena in Tissue Regeneration*. Baltimore: Williams & Wilkins Co.

LINDENBAUM, A. & KUETTNER, K. E. (1967). Mucopolysaccharides and mucoproteins of calf scapula. *Calc. Tiss. Res.* **1**, 153–65.

LINDHOLM, R., LINDHOLM, S., LIUKKO, P., PAASIMÄKI, J., ISOKÄÄNTÄ, S., ROSSI, R., AUTIO, E. & TAMMINEN, E. (1969). The mast cell as a component of callus in healing fractures. *J. Bone Jt Surg.* **51** B, 148–55.

LOPEZ, E. (1970). Demonstration of several forms of decalcification in bone of the Teleost fish, *Anguilla anguilla* L. *Calc. Tiss. Res.* **4** (suppl.), 83.

LÖRCHER, K., & NEWESELY, H. (1969). Calcium carbonate (calcite) as a separate phase besides calcium phosphate apatite in medullary bone of laying hens. *Calc. Tiss. Res.* **3**, 358–62.

MCCLURE, F. J. (1970). *Water Fluoridation. The Search and the Victory.* U. S. Dept. Health, Education and Welfare. Nat. Inst. Dent. Res., Bethesda, Maryland.

MCCLUSKEY, R. T. & THOMAS, L. (1958). The removal of cartilage matrix, *in vivo*, by papain. Identification of crystalline papain protease as the cause of the phenomenon. *J. exp. Med.* **108**, 371–84.

MACINTYRE, I. (1967). Calcitonin: a general review. *Calc. Tiss. Res.* **1**, 173–82.

MACK, P. B., LACHANCE, P. A., VOSE, G. P. & VOGT, F. B. (1967). Bone demineralization of foot and hand of Gemini–Titan IV, V and VII astronauts during orbital flight. *Am. J. Roentg.* **100**, 503–11.

MCLEAN, F. C. & BLOOM, W. (1940). Calcification and ossification. Calcification in normal growing bone. *Anat. Rec.* **78**, 333–59.

MCLEAN, F. C. & BLOOM, W. (1941). Calcification and ossification. Mobilisation of bone salt by parathyroid extract. *Arch. Path.* **32**, 315–33.

MCLEAN, F. C. & URIST, M. R. (1968). *Bone*, 3rd edn. Chicago: University of Chicago Press.

MCLOUGHLIN, C. B. (1968). Interaction of epidermis with various types of foreign mesenchyme. In *Epithelial-mesenchymal Interactions* (eds. R. Fleischmajer and R. E. Billingham), pp. 244–51. Baltimore: Williams & Wilkins Co.

MANN, G. (1902). *Physiological Histology. Methods and Theory.* Oxford: Clarendon Press.

MARINO, A. A. & BECKER, R. O. (1970). Evidence for epitaxy in the formation of collagen and apatite. *Nature, Lond.* **226**, 652–3.

MARKS, S. C. (1969). The thyroid parafollicular cell as the source of a potent osteoblast-stimulating factor: evidence from osteopetrotic mice. *J. Bone Jt Surg.* **51 A**, 875–90.

MARTIN, J. H. & MATTHEWS, J. L. (1969). Mitochondrial granules in chondrocytes. *Calc. Tiss. Res.* **3**, 184–93.

MATHEWS, M. B. (1967). Macromolecular evolution of connective tissue. *Biol. Revs.* **42**, 499–544.

MATHEWS, M. B. & LOZAITYTE, I. (1958). Sodium chondroitin-sulphate–protein complexes of cartilage. I. Molecular weight and shape. *Archs Biochem. Biophys.* **74**, 158–74.

MEACHIM, G. & CORNAH, M. S. (1970). Fine structure of juvenile nucleus pulposus. *J. Anat.* **107**, 337–50.

MELLORS, R. C. (1966). Electron microprobe analysis of human trabecular bone. *Clin. Orthop.* **45**, 157–67.

MILLER, E. J. & MARTIN, G. R. (1968). The collagen of bone. *Clin. Orthop.* **59**, 195–232.

MILLER, E. J., MARTIN, G. R., PIEZ, K. A. & POWERS, M. J. (1967). Characterisation of chick bone collagen and compositional changes associated with maturation. *J. biol. Chem.* **242**, 5481–9.

MINCHIN, E. A. (1898). Materials for a monograph of the ascons. I. On the origin and growth of the triradiate and quadriradiate spicules in the family Clathrinidae. *Q. J. microsc. Sci.* **40**, 469–587.

MINCHIN, E. A. (1908). Materials for a monograph of the ascons. II. The formation of spicules in the genus *Leucosolenia* with some notes on the histology of the sponges. *Q. J. microsc. Sci.* **41**, 301–55.

MOODIE, R. L. (1926). Studies in paleopathology, XIII. The elements of the Haversian system in normal and pathological structures among fossil vertebrates. *Biologia gen.* **2**, 63–95.

MOODIE, R. L. (1928). The histological nature of ossified tendons found in dinosaurs. *Am. Mus. Novit.* **311**, 1–15.

MOSCONA, A. (1952). Cell suspensions from organ rudiments of chick embryos. *Expl Cell Res.* **3**, 535–9.

MOSKALEWSKI, S. (1963). Studies on the osteogenic properties of uncultured and cultured isolated cells of the transitional epithelium. *Bull. Acad. pol. Sci.* **11**, 303–7.

MOSS, M. L. (1961 *a*). The initial phyllogenetic appearance of bone: an experimental hypothesis. *Trans. N.Y. acad. Sci.* **23**, 495–500.

MOSS, M. L. (1961 *b*). Studies of the acellular bone of teleost fish. I. Morphological and systematic variations. *Acta anat.* **46**, 343–62.

MOSS, M. L. (1962). Studies of the acellular bone of teleost fish. II. Response to fracture under normal and acalcaemic conditions. *Acta anat.* **48**, 46–60.

MOSS, M. L. (1963). The biology of acellular teleost bone. *Ann. N.Y. Acad. Sci.* **109**, 337–50.

MOSS, M. L. (1965). Studies of the acellular bone of teleost fish. *Acta anat.* **60**, 262–76.

MOSS, M. L. (1968). The origin of vertebrate calcified tissue. In *Current Problems of Lower Vertebrate Phylogeny* (ed. T. Ørvig), pp. 359–71. Stockholm: Almqvist & Wiksell.

MUIR, H., BULLOUGH, P. & MAROUDAS, A. (1970). The distribution of collagen in human articular cartilage with some of its physiological implications. *J. Bone Jt Surg.* **52** B, 554–63.

MUNSON, P. L. & HIRSCH, P. H. (1966). Thyrocalcitonin: newly recognized thyroid hormone concerned with metabolism of bone. *Clin. Orthop.* **49**, 209–32.

MURRAY, P. D. F. (1936). *Bones.* London: Cambridge University Press.

NEUMAN, W. F. & NEUMAN, M. W. (1958). *The Chemical Dynamics of Bone Mineral.* Chicago: University of Chicago Press.

NICHOLS, G. JR. (1970). Bone resorption and calcium homeostasis: one process or two? *Calc. Tiss. Res.* **4** (suppl.), 61–73.

NOBLE, H. W., CARMICHAEL, A. F. & RANKONE, D. M. (1962). Electron microscopy of human developing dentine. *Archs oral Biol.* **7**, 395–417.

ØRVIG, T. (1965). Palaeohistological notes. 2. Certain comments on the phyletic significance of acellular bone tissue in lower vertebrates. *Ark. Zool.* **16**, 551–6.

OSTROWSKI, K., RYMASZEWSKA, T., MOSKALEWSKI, S., WŁODARSKI, K. & ZALESKI, M. (1967). Experimental data on bone induction and bone antigenicity. *Symp. Biol. Hung.* **7**, 227–40.

OWEN, J. J. T. (1963). The mode of action of parathyroid hormone on bone. M.D. thesis, University of Liverpool.

OWEN, M. (1963). Cell population kinetics of an osteogenic tissue. *J. Cell Biol.* **19**, 19–32.

OWEN, M. (1967). Uptake of ^{3}H uridine into precursor pools and RNA in osteogenic cells. *J. Cell Sci.* **2**, 39–56.

OWEN, M. (1970). The origin of bone cells. *Int. Rev. Cytol.* **28**, 213–38.

OWEN, M. & BINGHAM, P. J. (1968). The effect of parathyroid extract on RNA synthesis in osteogenic cells *in vivo*. In *Parathyroid Hormone and Thyrocalcitonin (Calcitonin)* (eds. R. V. Talmage and L. F. Belanger), pp. 216–25. *International Congress Series*, no. 159. Amsterdam: Excerpta Medica Foundation.

OWEN, M. & SHETLAR, M. R. (1968). Uptake of ^{3}H-glucosamine by osteoclasts. *Nature, Lond.* **220**, 1335–6.

PAUTARD, F. (1961). Calcium, phosphorus and the origin of backbones. *New Scientist* **260**, 364–6.

PELLEGRINO, E. D. & BILTZ, R. M. (1968). Bone carbonate and the Ca to P molar ratio. *Nature, Lond.* **219**, 1261–2.

PETRUSKA, J. A. & HODGE, A. J. (1964). A sub-unit model for the tropocollagen macromolecule. *Proc. natn. Acad. Sci. U.S.A.* **51**, 871–6.

PIEZ, K. A., EIGNER, E. A. & LEWIS, M. S. (1963). The chromatographic separation and amino acid composition of the sub-units of several collagens. *Biochemistry, N.Y.* **2**, 58–66.

POSNER, A. S., PERLOFF, A. & DORIO, A. F. (1958). Refinement of the hydroxyapatite structure. *Acta cryst.* **11**, 308–9.

POST, R. W., HEIPLE, K. G., CHASE, S. W. & HERNDON, C. H. (1966). Bone grafts in diffusion chambers. *Clin. Orthop.* **44**, 265–70.

PRESCOT, G. H., MITCHELL, D. F. & FAHMY, H. (1968). Procion dyes as matrix markers in growing bone and teeth. *Am. J. phys. Anthropol.* **29**, 219–24.

PRITCHARD, J. J. (1964). Histology of fracture repair. In *Modern Trends in Orthopaedics. 4. Science of Fractures* (ed. J. M. D. Clarke), pp. 69–90. London: Butterworth.

PUTTE, VAN DE, K. A. & URIST, M. R. (1966). Experimental mineralisation of decalcified bone. *Clin. Orthop.* **44**, 273.

QUEKETT, J. (1846). On the intimate structure of bone. *Trans. microsc. Soc. Lond.* **2**, 46–58.

RAMACHANDRAN, G. N. (1963). Molecular structure of collagen. In *Int. Rev. Conn. Tiss. Res. 1* (ed. D. A. Hall), pp. 127–82. London and New York: Academic Press.

RAMACHANDRAN, G. N. (1968). Structure of collagen at the molecular level. In *Treatise on Collagen. I Chemistry* (ed. G. N. Ramachandran), pp. 103–83. London and New York: Academic Press.

RANVIER, L. (1889). *Traité Technique d'Histologie*, 2nd edn. Paris: Libraire Savy.

RASMUSSEN, H. & PECHET, M. A. (1970). Calcitonin. *Scientific American*, **223**, 4, 42–50.

RECKLINGHAUSEN, VON, F. (1910). *Untersuchungen über Rachitis und Osteomalacie*. Fischer: Jena.

ROBIN, CH. (1849). Sur l'existence de deux espèces nouvelles d'éléments anatomiques qui se trouvent dans le canal médullaire des os. *C. r. Séanc. Soc. Biol.* **1**, 149–50.

ROBINSON, R. A. & WATSON, M. L. (1955). Crystal–collagen relationships in bone as observed in the electron microscope. *Ann. N.Y. Acad. Sci.* **60**, 596–628.

ROBINSON, R. (1923). The possible significance of hexosephosphoric esters in ossification. *Biochem. J.* **17**, 286–93.

ROBINSON, R. (1932). *The Significance of Phosphoric Esters in Metabolism*. New York: University Press.

ROCKOFF, S. D. & ARMSTRONG, J. D. (1970). Parathyroid hormone as a stimulus to mast cell accumulation in bone. *Calc. Tiss. Res.* **5**, 49–55.

ROGERS, A. W. (1967). *Techniques of Autoradiography*. London: Elsevier.

ROSENBERG, L., HELLMAN, W. & KLEINSCHMIDT, A. K. (1970). Macromolecular models of protein–polysaccharides from bovine nasal cartilage based on electron microscope studies. *J. biol. Chem.* **245**, 4123–30.

ROSS, R. & BENDITT, E. P. (1962). Wound healing and collagen formation. *J. Cell Biol.* **15**, 99–115.

ROSS, R. & BENDITT, E. P. (1965). Wound healing and collagen formation. *J. Cell Biol.* **27**, 83–106.

Rossi, C. S. & Lehinger, A. L. (1964). Stoichiometry of respiratory stimulation, accumulation of Ca^{3+} and phosphate and oxidative phosphorylation in rat liver mitochondria. *J. biol. Chem.* **239**, 3971–80.

Runham, N. W., Thornton, P. R., Shaw, D. A. & Wayte, R. C. (1969). The mineralization and hardness of the radular teeth of the limpet *Patella vulgata* L. *Z. Zellforsch. mikrosk. Anat.* **99**, 608–26.

Russell, R. G. E. & Fleisch, H. (1970). Inorganic pyrophosphate and pyrophosphatases in calcification and calcium homeostasis. *Clin. Orthop.* **69**, 101–17.

Salpeter, M. M. (1968). H^3-proline incorporation into cartilage: electron microscope autoradiographic observations. *J. Morph.* **124**, 387–422.

Schaeffer, B. (1961). Differential ossification in the fishes. *Trans. N.Y. Acad. Sci.* **23**, 501–5.

Schenk, R. K., Muller, J., Zinkernagel, R. & Willenegger, H. (1970). Ultrastructure of normal and abnormal bone repair. *Calc. Tiss. Res.* **4** (suppl.), 110–11.

Schenk, R. S., Spiro, D. & Wiener, J. (1967). Cartilage resorption in the tibial epiphyseal plate of growing rats. *J. Cell Biol.* **34**, 275–91.

Schmitt, F. O., Gross, J. & Highberger, J. H. (1955). States of aggregation of collagen. *Symp. Soc. exp. Biol.* **9**, 148–62.

Schneider, B. J. (1968). Lead acetate on a vital marker for the analysis of bone growth. *Am. J. phys. Anthropol.* **29**, 197–200.

Scott, B. L. & Pease, D. C. (1956). Electron microscopy of the epiphyseal apparatus. *Anat. Rec.* **126**, 465–95.

Scott, B. L. (1967). The occurrence of specific cytoplasmic granules in the osteoclast. *J. Ultrastruct. Res.* **19**, 417–31.

Scowen, E. F. (1940). The study of bone marrow grafts *in vivo*. *British Empire Cancer Campaign, Ann. Rep.* **17**, 55–6.

Scowen, E. F. (1941). The study of bone marrow grafts *in vivo*. *British Empire Cancer Campaign, Ann. Rep.* **18**, 52–3.

Scowen, E. F. (1942). The study of bone marrow grafts. *British Empire Cancer Campaign, Ann. Rep.* **19**, 1.

Seipel, C. M. & Hammar, J. A. (1938). An English translation of Sandström's *Glandulae Parathyreoideae*. *Bull. Inst. History Med.* **6**, 179–222.

Severson, A. R. (1969). Mast cells in areas of experimental bone resorption and remodelling. *Br. J. exp. Path.* **50**, 17–21.

Shapiro, I. M. & Greenspan, J. S. (1969). Are mitochondria directly involved in biological mineralization? *Calc. Tiss. Res.* **3**, 100–102.

Sheldon, H. & Robinson, R. A. (1957). Electron microscope studies of crystal–collagen relationships in bone. *J. biophys. biochem. Cytol.* **3**, 1011–16.

Simkiss, K. (1970). Effect of acetazolamide on intracellular pH of avian shell gland. *J. Physiol., Lond.* **207**, 63–4.

Smith, D. S. (1968). *Insect Cells. Their Structure and Function.* Edinburgh: Oliver & Boyd.

Smith, H. (1953). *From Fish to Philosopher.* Boston: Little, Brown & Co.; Toronto: McClelland & Stewart.

Smith, J. W. & Frame, J. (1969). Observations on the collagen and protein-polysaccharide complex of rabbit corneal stroma. *J. Cell Sci.* **4**, 421–36.

Smith, J. W., Peters, T. J. & Serafini-Fracassini, A. (1967). Observations on the distribution of the protein–polysaccharide complex and collagen in bovine articular cartilage. *J. Cell Sci.* **2**, 129–36.

Smith, J. W. & Serafini-Fracassini, A. (1968). The distribution of the protein-polysaccharide complex in the nucleus pulposus matrix in young rabbits. *J. Cell Sci.* **3**, 33–40.

SOBEL, A. E., BURGER, M. & NOBEL, S. (1960). Mechanisms of nuclei formation in mineralising tissues. *Clin. Orthop.* **17**, 103–23.

SOLOMONS, C. C. & IRVING, J. T. (1958). Reactions of some hard and soft tissue collagens with 1-fluoro-2:4-dinitrobenzene. *Biochem. J.* **68**, 499–503.

SPILLANE, J. D. & WELLS, C. E. C. (1969). *Acrodystrophic Neuropathy*. London: Oxford University Press.

STRATES, B. S. & URIST, M. R. (1969). Calcification in implants on tendon. *Experientia*, **25**, 924–6.

SUZUKI, H. K. (1963). Studies on the osseous system of the slider turtle. *Ann. N.Y. Acad. Sci.* **109**, 351–410.

TALMAGE, R. V. (1967). A study of the effect of parathyroid hormone on bone remodelling and on calcium homeostasis. *Clin. Orthop.* **54**, 163–73.

TARLO, L. B. H. (1963). Aspidin: the precursor of bone. *Nature, Lond.* **199**, 46–8.

TARLO, L. B. H. (1964). The origin of bone. In *Bone and Tooth Symposium* (ed. H. J. J. Blackwood), pp. 3–17. London: Pergamon Press.

TARLO, L. B. H. (1967). Biochemical evolution and the fossil record. A symposium with documentation. *Proc. geol. Soc. Lond.* **123**, 119–32.

TARLO, L. B. H. & MERCER, J. R. (1961). A note on the histological study of fossil dentine. *Proc. geol. Soc. Lond.* **1590**, 127–8.

TARLO, L. B. H. & MERCER, J. R. (1966). Decalcified fossil dentine. *J. Roy. microsc. Soc.* **86**, 137–40.

TAVES, D. R. (1965). Mechanisms of calcification. *Clin. Orthop.* **42**, 207–20.

TAYLOR, T. G. (1970). How an eggshell is made. *Scientific American* **222**, 3, 88–97.

TAYLOR, T. G. & BELANGER, L. F. (1969). The mechanism of bone resorption in laying hens. *Calc. Tiss. Res.* **4**, 162–73.

TERMINE, J. D. (1966). Amorphous calcium phosphate: the second mineral of bone. Thesis, Cornell University.

TERMINE, J. D. & POSNER, A. S. (1966). Infra-red analysis of bone. Age dependency of amorphous and crystalline mineral fractions. *Science, N.Y.* **153**, 1523–5.

THOMPSON, J. & URIST, M. R. (1970). Influence of cortisone and calcitonin on bone morphogenesis. *Clin. Orthop.* **71**, 253–70.

TODD, E. & BOWMAN, J. A. (1845). The *Physiological Anatomy and Physiology of Man*. Philadelphia: Blanchard & Lea.

TOMES, J. & DE MORGAN, C. (1853). Observations on the structure and development of bone. *Phil. Trans. Roy. Soc.* **143**, 109–39.

TRUETA, J. (1963). The role of the vessels in osteogenesis. *J. Bone Jt Surg.* **45**B, 402–18.

URIST, M. R., DOWELL, T. A., HAY, P. H. & STRATES, B. S. (1968). Inductive substrates for bone formation. *Clin. Orthop.* **59**, 59–95.

URIST, M. R., HAY, P. H., DUBUC, F. & BÜRING, K. (1969). Osteogenic competence. *Clin. Orthop.* **64**, 194–220.

URIST, M. R. & MCLEAN, F. C. (1941). Calcification and ossification. 1. Calcification in the callus in healing fractures in normal rats. *J. Bone Jt Surg.* **23**A, 1–16.

URIST, M. R. & NOGAMI, H. (1970). Morphogenetic substratum for differentiation of cartilage in tissue culture. *Nature, Lond.* **225**, 1051–2.

URIST, M. R., SILVERMAN, B. F., BÜRING, K., DUBUC, F. L. & ROSENBERG, J. M. (1967). The bone induction principle. *Clin. Orthop.* **53**, 243–83.

VAES, G. (1965). Hydrolytic enzymes and lysosomes in bone cells. In *Proceedings 2nd European Symposium on Calcified Tissues* (eds. I. C. Richelle and M. J. Dallemagne), pp. 51–62. Liege.

VAES, G. (1968). On the mechanisms of bone resorption. *J. Cell Biol.* **39**, 676–97.

VAES, G. (1969). Lysosomes and the cellular physiology of bone resorption. In *Lysosomes in Biology and Pathology*, vol. 1 (eds. J. T. Dingle and H. B. Fell). Amsterdam and London: North-Holland Publishing Co.

VAES, G. & JAQUES, P. (1965). Studies on bone enzymes. The assay of acid hydrolases and other enzymes in bone tissue. *Biochem. J.* **97**, 380–92.

VASINGTON, F. D. & MURPHY, J. V. (1962). Ca uptake by rat kidney mitochondria and its dependence on respiration and phosphorylation. *J. biol. Chem.* **237**, 2670–7.

VAUGHAN, J. M. (1970). *The Physiology of Bone*. Oxford: Clarendon Press.

VAUGHAN, J. M. & WILLIAMSON, M. (1964). Localisation of transuranic elements on bone surfaces. In *Bone and Tooth Symposium* (ed. H. J. J. Blackwood), pp. 71–83. London: Pergamon Press.

VEIS, A. & ANESEY, J. (1965). Modes of intermolecular cross-linking in mature insoluble collagen. *J. biol. Chem.* **240**, 3899–908.

VILMANN, H. (1969). The *in vivo* staining of bone with alizarin red S. *J. Anat.* **105**, 533–45.

VINCENT, J. (1957). Les remaniements de l'os compact marqué à l'aide de plomb. *Rev. belg. Path.* **26**, 161–8.

WARNER, S. P. (1964). Hydrolytic enzymes in osteoclasts cultured *in vitro*. *J. Roy. microsc. Soc.* **83**, 397–403.

WASE, A. W., SOLEWSKI, J., RICKES, E., & SEIDENBERG, J. (1967). Action of thyrocalcitonin on bone. *Nature, Lond.* **214**, 388–9.

WEATHERELL, J. A. & WEIDMANN, S. M. (1963). The distribution of organically bound sulphate in bone and cartilage during calcification. *Biochem. J.* **89**, 265–7.

WEIDENREICH, F. (1930). Das Knochengewebe. In *Handb. d. mikros. Anat des. Menschen* (edv Möllendorf), **2**, pt. 2, 391–623. Berlin: Springer.

WEIDMANN, S. M. (1963). Calcification of skeletal tissues. In *Int. Rev. Conn. Tiss. Res. 1* (ed. D. A. Hall), pp. 339–77. London and New York: Academic Press.

WHITEHEAD, R. G. & WEIDMANN, S. M. (1959). Oxidative enzyme systems in ossifying cartilage. *Biochem. J.* **72**, 667–72.

WILLIAMSON, M. & VAUGHAN, J. (1964). A preliminary report on the sites of deposition of Y, Am and Pu in cortical bone and in the region of the epiphyseal cartilage plate. In *Bone and Tooth* (ed. H. J. J. Blackwood), pp. 71–83. London: Pergamon Press.

WILLMER, E. N. (1965). The origins of collagen. In *Structure and Function of Connective and Skeletal Tissue* (eds. S. Fitton Jackson, R. D. Harkness, S. M. Partridge and G. R. Tristram), pp. 196–206. London: Butterworth.

WILLMER, E. N. (1970). *Cytology and Evolution*, 2nd edn. New York: Academic Press.

WINAND, L. & DALLEMANGE, M. J. (1962). Hydrogen bonding in the calcium phosphates. *Nature, Lond.* **193**, 369–70.

WINTER, G. D. (1970). Heterotopic bone formation in a synthetic sponge. *Proc. Roy. Soc. Med.* **63**, 1111–15.

WINTER, G. D. & SIMPSON, B. J. (1969). Heterotopic bone formed in synthetic sponge in the skin of young pigs. *Nature, Lond.* **223**, 88–90.

WŁODARSKI, K. (1970). The inductive properties of epithelial established cell lines. *Exptl Cell Res.* **57**, 446–8.

WŁODARSKI, K., HINEK, A. & OSTROWSKI, K. (1970). Investigations on cartilage and bone induction in mice grafted with FL and WISH line human amnion cells. *Calc. Tiss. Res.* **5**, 70–9.

WYCKOFF, R. W. G. & DOBERENZ, A. R. (1965). The electron microscopy of Rancho la Brea bone. *Proc. natn. Acad. Sci. U.S.A.* **53**, 230–3.

WYCKOFF, R. W. G., WAGNER, E., MATTER, P. & DOBERENZ, A. R. (1963). Collagen in fossil bone. *Proc. natn. Acad. Sci. U.S.A.* **50**, 215–18.

YOUNG, R. W. (1962*a*). Cell proliferation and specialization during endochondral osteogenesis in young rats. *J. Cell Biol.* **14**, 357–70.

YOUNG, R. W. (1962*b*). Autoradiographic studies on postnatal growth of the skull in young rats injected with tritiated glycine. *Anat. Rec.* **143**, 1–13.

YOUNG, R. W. (1963). Histophysical studies in bone cells and bone resorption. In *Mechanisms of Hard Tissue Destruction* (ed. R. F. Sognnaes). Washington: Am. Ass. Adv. Sci. publ. no. 75.

YOUNG, R. W. (1964). Specialization of bone cells. In *Bone Biodynamics*. (ed. H. M. Frost), pp. 117–37. Boston: Little, Brown & Co.

INDEX

absorption of bone, *see* resorption
acid phosphatase, in osteoclasts, 130
acrodystrophic neuropathy, 136
adenosine triphosphate (ATP), in calcifying cartilage, 88–9
age
  and arrangement of cellulose fibres in *Valonia*, 108
  and bonds in collagen chains, 39
  and content of ground substance in bone, 43
  and size of bone crystals, 47
alanine, in collagen, 37
Algire chambers, for study of osteo-induction, 164–5, 168, 172
ameloblasts (enamel formers), fig. 42
americium, concentrated in bone matrix, 4, fig. 2
amino acids
  of calcitonin, 175
  of collagen, 36–7; in calcification, 84, 86
  in fossil bone, 51–2
  labelled, for studying syntheses: by odontoblasts, 95; by osteoblasts, 75, 76, 77, 138, 139; by osteocytes, 103
aneurysm of artery, resorption of bone caused by, 113
apatites, 46
aspidin, in fossil dermal plates, 54, 55, 56, figs. 25, 27
aspidinoblasts, 54, 55
aspidinoclasts, 54
aspidinocytes, 55, figs. 26, 27 *a*, 28 *a*
aspidones, 54
astronauts, bone metabolism of, 3, 154
aureomycin, as tracer for bone formation, 109
autoradiography, 13–15, figs. 2, 3, 41
  of odontoblasts, 95
  of osteoblasts, 75–6
  of osteoclasts, 124, 138–9
  of tissue implants, 171

barium, replacing calcium in bone, 46
bats, non-vascularized bone in, 58
bicarbonate, replacing hydroxyl in bone, 45–6
birefringence, of bone lamellae, figs. 13, 15

blood calcium
  calcitonin and, 4, 175
  osteocytes and, 150
  parathyroid hormone and, 3, 136, 137, 139, 142
blood vessels, *see* vascular channels
bone marrow, 3, 4, 28, 114, figs. 4, 16, 56
  induction of bone by, 3, 169
  tumour cells invading, figs. 57, 58
bone mineral, 3, 18, 44–9, figs. 32, 39, 40
  non-crystalline in earliest stage, 48, 83, 84, 89–90
  in osteoclasts, 123, 127, 130, fig. 67*b*
bone modelling (remodelling), 67, 113, 159, fig. 12
  in fossil dermal plates, 54
bone turnover, 24, 26, 35, 41, 108, 152

calcifiability of collagen, 58, 84
calcification, 78, 84–92, figs. 39, 40
  different densities of, 12, 45, fig. 1
  mitochondria and, 70, 79–81
  techniques for study of, 82–4
  of various tissues, in osteitis fibrosa, 142
Calcified Tissue Research, 5
calcite (calcium carbonate)
  in bone, 48, 49
  in spicules of calcareous sponges, 81, 96
calcitonin, 3–4, 175–6
  excess of parathyroid hormone and, 143, 144
calcium
  accumulated by mitochondria, 79–80
  bound by sialoprotein of ground substance, 87
  calcitonin and metabolism of, 4, 175
  ions of: in bone, 45, 46, 82; in calcification, 86, 87, 88, 89, 90
  osteocytes and blood level of, 150
  parathyroid and metabolism of, 3, 136, 137, 139, 142
  ratio of phosphorus to, and crystal size of apatite, 47
calcium phosphate (calcium hydroxyapatite), 3, 18, 45–9, 93
calcoblasts of sponges, 16, 81, 96–9, figs. 44, 45
callus, *see* fracture callus

Date Due

MY 7 '78			
FE 28 76			
NOV 3 '83			
ILL			
5-11-94			
DEC 15 1997			

Demco 38-297